Urban Astro

7/2004

Light pollution has spread so much in the last few decades that it compromises our view of the stars. It is becoming more and more difficult to find an observing site with clear, dark skies away from light and industrial pollution. However, with patience and some simple equipment, and by choosing the right target to observe, amateur astronomers can still find observing from towns and cities a rewarding hobby.

The result of thirty years of observing the night sky from the city, Denis Berthier's practical guide will help amateur astronomers to enjoy their hobby without having to travel to distant sites, and without using complicated equipment or difficult techniques. It will enable them to observe and photograph stars, planets and other celestial objects from their own town.

Denis Berthier is a French journalist who has been passionate about astronomy for the last thirty years. He has been Laureate of the French Association for Astronomy, and has published numerous papers on astronomical photography, and instrument construction. He has written a previous book with Jean Lacroux, *Lunettes et telescopes, mode d'emploi*, published by Bordas.

Urban
Astronomy

Denis BERTHIER
(translated by Klaus Brasch)

CAMBRIDGE
UNIVERSITY PRESS

PUBLISHED BY THE PRESS SYNDICATE OF THE UNIVERSITY OF CAMBRIDGE
The Pitt Building, Trumpington Street, Cambridge, United Kingdom

CAMBRIDGE UNIVERSITY PRESS
The Edinburgh Building, Cambridge CB2 2RU, UK
40 West 20th Street, New York, NY 10011–4211, USA
477 Williamstown Road, Port Melbourne, VIC 3207, Australia
Ruiz de Alarcón 13, 28014 Madrid, Spain
Dock House, The Waterfront, Cape Town 8001, South Africa

http://www.cambridge.org

French edition © Bordas/HER, 2001
This edition © Cambridge University Press, 2003
Published with the help of the French Ministry of Culture

Previously published in French
Observer le ciel en ville,
by Denis Berthier 2001
First published 2003

Printed in the United Kingdom at the University Press, Cambridge

Typeface: Formata and Garamond BE 10.5/12pt.   System: QuarkXpress™ [SE]

A catalogue record for this book is available from the British Library

Library of Congress Cataloguing in Publication data

Berthier, Denis, 1949–
  [Observer le ciel en ville. English]
  Urban astronomy / Denis Berthier; translated by Klaus Brasch.
    p. cm.
  Includes bibliographical references and index.
  ISBN 0-521-53190-X
  1. Astronomy – Observers' manuals.   I. Title.

  QB64.B4913 2003
  522′.09173′2–dc21   2002035185

ISBN 0 521 53190 X paperback

Contents

Foreword

At first glance, doing astronomy from the city seems like a totally incongruous activity and indeed many amateur astronomers have become quite frustrated trying to observe from urban locations. Light pollution, smog, haze, and dust, have all combined to not only dim the sky above us but also to dampen our enthusiasm to observe it. Urban astronomy is certainly not an easy thing to do.

Nevertheless, far more amateurs than one would imagine regularly observe, photograph, and enjoy the stars from their backyards, driveways, balconies, and rooftops. This book is the result of over 30 years experience observing from urban settings. It provides both beginners and more advanced amateurs advice and many hints on how to best observe under these conditions.

The book provides a practical, hands-on approach to astronomy, as well as advice on what equipment and conditions are best suited for observing stars, planets, and other astronomical objects of interest. With a little bit of patience and some experience, the urban astronomer can continue to observe and enjoy the universe around him.

Denis Berthier

Getting started

Solid footing and good viewing

The sky is clear this evening, the barometer is rising, the weather forecast is promising, and you have decided finally to go out and do some serious stargazing. What equipment should you take and from what location will you observe? Keep it simple and for starters use your own eyes.

Setting up your observing site

Before anything else, the urban astronomer must improvise a suitable "mini-observatory". Ideally this would be a dark corner of your garden or back yard, shielded from direct light and showing at least a quarter of the sky toward the south or southwest. Failing that, a balcony, driveway, or patio with similar orientation will do. The drawback there is that cement or concrete floors absorb sunlight during the day and radiate it all back at night. This can cause air turbulence and degrade telescopic images, although it will not affect naked eye observing.

Even if your only access to the sky is through a window, do not worry; you can still observe many interesting astronomical objects. The main thing to remember is to shield your eyes and line of sight from artificial light as much as possible. Turn off your own house lights and try to block off neighborhood lights with anything you can, makeshift screens, an umbrella or tarp, whatever works. Also, do not hesitate to ask your neighbors to lower that porch light or even turn it off completely for a while; people can be quite accommodating in that regard once you explain why.

Start with the naked eye

Your eyes are extraordinarily sensitive instruments for nighttime viewing and, once fully dark adapted, are many times more sensitive at night than during the day. This is in part because your pupils dilate in the dark to let more light in. In addition, the retina also has two types of photo-receptive cells, rods and cones. Both types transmit light signals directly to the brain; however, only cones detect colors, while the rods are sensitive to the intensity of light. Color perception decreases at night, but the rods become more sensitive. That explains why we do not see colors well at night and why things always appear monochromatic by moonlight. To enjoy the full benefit of your nighttime visual acuity, it is crucial that your eyes have had at least 20 minutes to become fully dark adapted. During that time the rods rapidly accumulate more light-sensitive pigment, allowing you to see increasingly fainter objects

A patio or a balcony can serve as your observatory. A balcony is particularly well suited for small instruments like this 70-mm refractor.

and detail. You will be astonished by how many more stars you will be able to see even from the city

To help you see better in the dark and also let you read star charts without destroying your "night vision," get yourself a pocket flash light with a dim, red colored bulb or covered with red cellophane or transparent wrapping paper.

Keeping track of your observations

An indispensable accessory for any observer is a good notebook. Use it to record and log everything in writing and through sketches, including dates, times, location, equipment used, and any special comments or reactions. You will find this both instructive and enjoyable, especially some time later as you recall your experiences and follow your progress. Occasionally too, such notes may help in case you have made a significant discovery or novel observation.

Drawing what you observe can also be very satisfying for you, especially as you follow the progress of a lunar eclipse or some other astronomical phenomenon. Well-known astronomers, including Galileo, Camille Flammarion, and many others routinely recorded and sketched what they saw and observed.

Using the Sun to orient yourself

If you are just beginning and are not sure which way you are facing, north, west, south, etc., or whether your planned observing site is really the best location for stargazing, orient yourself first during the day. The Sun can be very helpful in that regard, since it rises in the east, passes the meridian at midday, and sets in the west. As the seasons change, you will note progressive changes too in the path of the Sun across the sky. In summer it rises more toward the northeast

The sun sets in the west and rises in the east. Remember these positions with reference to local landmarks (buildings, trees, antennas).

and sets in the northwest. In winter when days are shorter, the Sun rises in distinctly more southeasterly directions and also sets in the southwest. A compass can be very helpful here, although with a little practice you will quickly learn the cardinal points and your orientation relative to the position of the Sun and the time of day. Remember to make adjustments for local time during the summer, when solar midday is closer to 2 p.m. (14:00 hrs).

Once you become familiar with the major constellations, seek out your favorite stars to help orient yourself at night, just like sailors have done for centuries. Like the Sun the stars and constellations also appear to rise in the east, move along the meridian and overhead (zenith) and set in the west. The stars of course do not literally move across the sky, their actual or proper motion is much too slow to be seen with the naked eye. The daily apparent motion of the Sun, stars, and constellations across the sky is simply a reflection of the Earth's counterclockwise rotation, from west to east. In sum, by properly orienting your mini-observatory during the day, you will be better oriented for nighttime observing as well, particularly if you can face toward the south or southwest.

An initial tour of the sky

*W*hat is the nature of those intriguing and scintillating points of light in the sky? This question was of great concern to our ancestors and the earliest observers. By combining religious beliefs, mythology, and rudimentary science, ancient cultures assigned names, legends, and associations to the stars above.

The birth of constellations

The observers of antiquity saw imaginary outlines of heros and animals in the sky (Hercules, lions, etc.) and so the constellations, geometric asterisms with exotic names were born and survive to this day. Letters of the Greek alphabet and their stars designate the official names of constellations. The names of most principal stars are in Arabic however, since during the Middle Ages Arab astronomers first catalogued them.

The outlines of constellations are quite arbitrary since the stars in them have no actual physical connection but only appear associated in the sky by chance and line of sight. Modern astronomy recognizes 88 official constellation groups in the sky but only about 50 of these are visible from mid-northern locations, like France.

The Big Dipper: key to the northern sky

We learned before that the Sun sets directly in the west during the spring and autumn. Make note of those directions. After sunset, make a quarter turn to the right and wait for darkness to set in. You are now facing directly north. Look up and you will notice seven brilliant stars in the shape of a ladle or Big Dipper, also known officially as Ursa Major, the big bear.

If you observe the Big Dipper at the same time on successive evenings throughout the

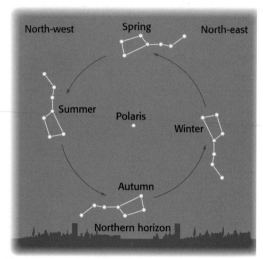

The Big Dipper's seasonal rotation cycle in the sky.

year, you will notices its changing seasonal orientation. During the spring, it will be straight overhead at the zenith around 11 p.m. About the same time in the summer, it will appear above the western horizon. In the autumn, you can spot it just above the northern horizon, and, in winter, near the eastern horizon.

These seasonal changes in the Big Dipper's position reflect changes in the Earth's orientation as it revolves around the Sun. In a similar way, the apparent counterclockwise motion of this constellation during a single night, reveals the Earth's rotation around its own axis. The Big Dipper and several other constellations close to the North celestial pole, never seem to set at mid northern latitudes, and are also known as "circumpolar" constellations.

Polaris, the centre of the northern sky

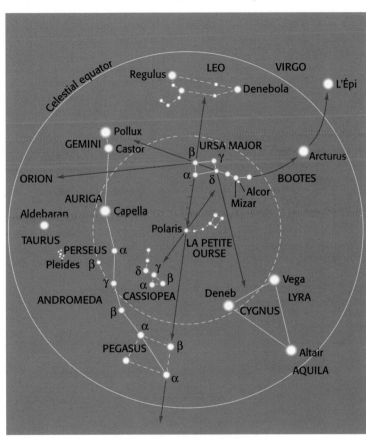

Pick out Mérak (β) and Dubhe (α), the two bright stars on the short side of the Big Dipper, and project a distance from β to α about five times their separation. You will discover a moderately bright star of 2nd magnitude at that point, situated very close to the North celestial pole. This is Polaris, the North Star, the central point around which everything else seems to rotate over a period of about 24 hours (or 23 hours, 56 minutes

and 18 seconds, to be precise, the exact duration of the sidereal day).

Polaris is the tip of the tail of little bear, also known as Ursa Minor or the Little Dipper. From the city you will probably spot only Kochab (β) and Pherkad (γ), the two end stars of the Little Dipper, since they are only 2nd and 3rd magnitudes, respectively. If you extend an imaginary line from Megrez (δ) in the Big Dipper, through Polaris and an equal distance beyond that, you will note a big "W" or "M" (depending on the time of year). This is the constellation Cassiopeia. Moving in the same general direction, you will come to the Andromeda and the great square of Pegasus

Take a good look at the accompanying star chart to more fully appreciate this long-standing method of sky orientation first practiced by ancient mariners. If you extend another line along the handle of the Big Dipper, you will inevitably spot brilliant Arcturus in the constellation of Bootes, followed by Spica in Virgo. By extending a line from Megrez (δ) through Merak (β) in the Big Dipper, you will move in the direction of Castor and Pollux in the constellation Gemini. By joining Phekda (γ) with Megrez (δ), you will reach the "Great Summer Triangle" (described more later), with the famous three stars Vega, Deneb, and Altair. The great Megrez–Merak axis also points to Orion the hunter.

The sky à la carte

Finding your way around the sky without a star map is like a trip into unknown territory. You risk losing yourself along a path with countless fascinating sites and locations.

A worthwhile investment

A planisphere or star wheel should be your first observing instrument. Serving both as star map and celestial calendar, it lets you locate the constellations very quickly at any time of year or night. A point to remember, most of these devices are calibrated to work properly in Universal Time (UT), the mean solar time of the Greenwich (England) meridian. Remember that in the sky designations like "summer" or daylight saving time or wintertime have little or no relevance, and you must adjust your planisphere accordingly to get the most accurate map of the sky at a particular time or season. In more advanced star wheels, a graduated, clear plastic cover centered on Polaris allows you to read the celestial coordinates of any star in the sky, namely its right ascension and declination, corresponding to terrestrial latitude and longitude.

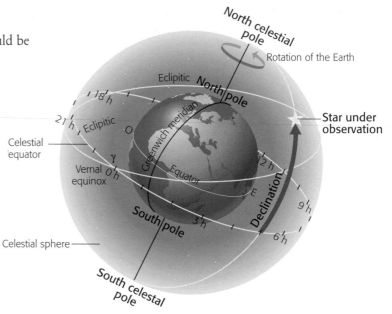

The principal celestial coordinates are similar to those used on the Earth: longitude and latitude become right ascension and declination. Once those two coordinates are given, any object in the sky can be found.

It is easy to use

Once you have adjusted the date and time on your planisphere, hold it up at an angle and in a way that coincides with the four cardinal points. For example, the planisphere's western horizon should **point down** if you are facing west, similarly its **northern horizon** should **point down** if you are facing north, and so on. As a general rule, any star chart should be oriented in such a way that the section of the horizon under observation points down toward the cardinal point you are facing.

READING THE SKY

Although a huge selection of star wheels is available, their basic design and operating principles are similar: most consist of a fixed all-sky star map overlain with a movable circle outlining the horizon and section of sky typically visible at any given time from mid-northen latitudes. The edge of the star wheel will typically indicate the months and days of the year. The movable section will be calibrated into 24 hours and also be marked to show two- or five-minute intervals. All you have to do is align the time of night with the month and day you plan on observing and you will see at a glance what section of the sky lies overhead.

Star wheel

Movable ruler

Movable hour circle

Fixed outer circle indicating months and day

Locating planets and stars

Some commercial planispheres are supplied with planetary ephemeredes, but, if not, you can always obtain this information from magazines, manuals, and several Web sites (see references). Ephemeredes list the right ascension and declination of the principal planets at 15-day intervals throughout the year. These coordinates should be shown on your star chart. Better-quality planispheres have a graduated, rotatable ruler for this purpose. **Right ascension** is read from the perimeter of the star wheel and the **declination** with the movable ruler. This may seem a little complicated at first, but once you try it a few times it is no more difficult than setting your watch. Although you may not be able to see many stars from the city, a planisphere will really help orient yourself and locate the brightest stars and major constellations.

Understanding the universe

Lost in some undefined portion of the universe and containing several trillion stars, our galaxy can be compared to a single grain of sand at the beach. The universe itself is populated with more galaxies than there are grains of sand on that beach.

Suns by the trillions

It is really quite difficult to fully appreciate the size and shape of the universe, particularly from deep within the city, where excessive amounts of artificial light blots out all but a few dozen stars. And yet our galaxy is estimated to contain several trillion stars, each of which is a sun not unlike ours, many existing in pairs or larger groupings. While the list of stars now known to have planetary companions grows almost weekly, it seems unlikely that double or multiple stars have developed planetary systems like those of our own Sun.

Stars associate by the trillions into enormous structures called galaxies. Most galaxies, or "island universes" as they were first termed in the 19th century, are disk or saucer shaped, with a central bulge from which spiral arms extend outward. There are also many galaxies with less-defined shapes and which have structures that are quite irregular in appearance.

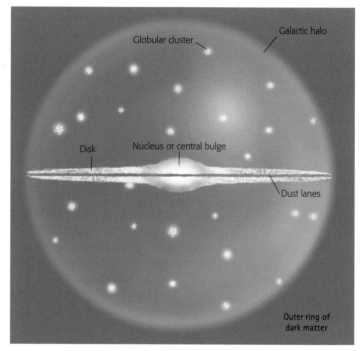

Globular cluster

Galactic halo

Disk

Nucleus or central bulge

Dust lanes

Outer ring of dark matter

This is roughly what our galaxy looks like as seen from a distance. Our Sun is located about a third the distance from the center to the edge.

CAN WE SEE A GALAXY FROM THE CITY?

Yes, you certainly can. Some of the brighter galaxies are observable from the middle of the city. Among these, the Andromeda galaxy (M31), a sister galaxy to our own, is clearly visible in binoculars although it is located some 2.2 million light years from us. Just think, the image you perceive in your binoculars or telescope left this galaxy about 2 200 000 years ago, a time when the first humans roamed the plains of Africa.

The Andromeda galaxy, a close neighbor, is very similar to our own Milky Way galaxy and contains an estimated 100 billion suns.

All the stars one can see, be it from the city or under dark skies, are part of our own galaxy, also known as the Milky Way. The Milky Way is best seen in the summer when it appears as a faint, glowing band that stretches across the entire sky. This appearance is due to the enormous number of stars in our galaxy that are too far away to be distinguished individually by eye.

Clusters and superclusters

Most galaxies are associated into groups or clusters with each individual member galaxy separated by several million light years. Our Milky Way galaxy is a member of the local group, which also includes the Virgo cluster of galaxies, a group of galaxies located in the direction of the constellation Virgo.

The next levels in this galactic hierarchy are even larger associations of clusters known as "super-clusters," whose dimensions and distances are expressed in "gigaparsecs" (see below). It is not yet known what the highest level of organization is in this hierarchy and what its limits might be within the observable universe.

Measuring interstellar distances

When measuring or expressing interplanetary and interstellar distances, there is little point in using conventional yardsticks like miles or kilometers. Even when expressing relatively "short" distances astronomers use measures like the "Astronomical Unit (AU)," the mean distance between the Earth and the Sun, or approximately 150 million km (96 million miles). While the AU is a suitable benchmark measure within the solar system, it is quite inadequate for the much greater distances beyond, that is where distances measured in "light years" (LY) apply, or the distance traveled by light in one year. Since the speed of light is approximately 300,000 km/sec. (186,000 miles/sec), one light year equates to a distance of nearly 10 trillion km (or 5.8 trillion miles). Vega, the major star in Lyra is 26 LY distant. Another standard astronomical yardstick is the parsec. This corresponds to the distance at which an imaginary star measured from Earth exhibits a 1 arc second angle of parallax. One parsec is equal to 3.26 LY, and interstellar distances are usually stated in kilo-parsecs (1000 parsecs).

The life and death of stars

Have you ever witnessed the birth of a star or the dramatic end of one? Stars are born, live and die by chemical and thermonuclear processes and under the universal influences of gravity. Astronomers have long followed and studied these phenomena with great interest.

Star birth

The space between the stars (interstellar space) is vast but certainly not totally empty. In addition to stars, planets and other gravitationally linked companions, interstellar space contains diffuse, but immense clouds composed of mostly hydrogen and microscopic dust particles. Hydrogen is the most abundant and most ancient element in the universe. It is also the simplest element, consisting of an atomic nucleus with a single proton and a single orbital electron. This simple element is the basic building block which gave rise to all the galaxies and stars.

The influences of the force of gravity pervade the entire universe. In short, all objects traveling in space are subject to gravitational interactions, which, sooner or later, bring them into close physical proximity. The giant clouds or nebulae of interstellar gas and dust are also subject to such interactions. Eventually this induces localized gravitational contraction in the nebula and formation of accretion globules or centers of star formation. This causes a dramatic increase in temperature and accelerates gravitational contraction to a point where the globules begin to spin or rotate, leading to formations of "proto-stars." These may form singly or in groups of two or more.

Eventually the density of matter in the protostar increases under gravitational contraction to the point where nuclear fusion begins (not to be confused with "nuclear

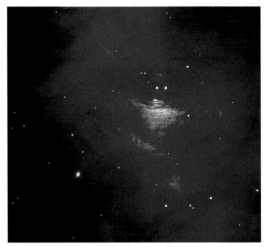

The famous Crab Nebula, the residue of a supernova explosion observed in 1054 and containing a pulsar at its core.

A STAR IN FULL DAYLIGHT

On July 4, 1054, somewhere in China, the Imperial Astrologer witnessed a most extraordinary spectacle, a star four times as bright as the full moon was visible in full daylight even at noon. The astonishing brightness of this "visitor" persisted for several days before gradually dimming until several weeks later it was no longer visible. Today, a powerful telescope is needed to see the remnants of supernova 1054, the Crab nebula in the constellation Taurus. Without realizing it, the Chinese observers had witnessed the "swan song" of a star in the last stages of its life. Stars are born, live, and die, sometimes in truly spectacular fashion . . .

The great Orion Nebula, a veritable nursery of very young stars about 1500 light years from us.

fission" as in an atomic bomb), when four atoms of hydrogen combine to form a new element, helium, containing two protons, two neutrons, and two electrons. This reaction liberates a phenomenal amount of energy, leading to an enormous increase in temperature and radiation. If the initial mass of the pre-stellar gas cloud is large enough, a luminous star is formed or "ignited," with internal temperatures rapidly approaching several million degrees at the center. A residual ring of gas and dust that may eventually condense into planets may surround the new, young star. That is what happened in the case of our own Sun, which is accompanied by a system of nine planets.

An energy powerhouse

For their entire existence, new stars have basically a single function: to convert hydrogen into helium, leading to release of light and heat energy and the synthesis of trace amounts of carbon, iron, and other metals. This is exactly what our Sun has been doing for nearly 5 billion years, while consuming some 4 million tons of hydrogen in the process every second. Clearly there is little reason for concern that the Sun will run out of hydrogen anytime soon, since stars are veritable H-bombs, with plenty of energy in store. The *total* estimated lifetime of the Sun is about 10–12 billion years.

Our Sun is a small to medium star, with a diameter of 1.39 million km or 109 times the diameter of the Earth. Some stars are significantly larger and more massive. Red giants like Betelgeuse in the constellation of Orion are less dense and could easily encompass the Earth's entire orbit. Antares in Scorpius is 480 times the diameter of the Sun and the orbit of Mars could fit within it, while Epsilon Aurigae could contain the entire Solar system!

A variety of colors

Once you become really familiar with the constellations you will notice that not all stars are white, but display a variety of distinctive colors, from red to blue, yellow, and green. These colors reflect the surface temperatures of stars as well as their ages. Blue indicates young, very hot stars, yellow–green are mid range temperature stars, near the middle of their life cycles. Our Sun is in that category. Red and orange stars are relatively cooler,

The Helix Nebula in Aquarius is evidence of a red giant that has shed most of its mass into space. The faint star in the center of this nebula is all that remains: a white dwarf.

older stars, slowly evolving toward the end of their lives as novas.

Stars are usually born in groups of several dozen members, which remain in close associations known as **open clusters**. Examples of these include the Haydes and Pleiades in Taurus, young clusters of stars less than 200 million years old and splendid objects to observe with binoculars. Our galaxy is also surrounded by a giant halo of much more compact and dense clusters that are composed primarily of very old stars. These stellar "retirement homes," containing several hundred thousand members, are know as **globular clusters**, because of their round or spherical shapes. The best-known globular cluster (in the northern hemisphere) is M 13 in Hercules.

A state of crisis

Depending on their mass and size, stars can live from several hundred million to several billion years. After this time, when they have converted almost all their hydrogen to helium, they enter a stage of disequilibrium leading rapidly to a state of crisis. Lacking sufficient hydrogen "fuel," they are now subject to forces that are no longer in balance, the outward pressure of expansion and the contracting force of gravity. Two scenarios are possible at this stage. For medium-size stars like our Sun, once hydrogen is used up, the core contracts causing the temperature to rise and the star expands to enormous proportions until it becomes a **red giant**. Such giants will slowly evolve and end their lives without much fuss by progressively shedding their mass into space.

The latter process can take several thousands of years. These moribund stars continue to throw off an expanding shell of gas, eventually becoming "planetary nebulae," so named because many appear as planet-like disks as seen in the telescope. At the center of such nebulae, one can usually detect a faint, bluish stellar residue, which is relatively dense and not much larger than the Earth. One liter of material from a star of this type would weigh about 2000 tons, the weight of about 100 train engines.

The power of gravity

Large, really massive stars face a far more catastrophic end of life. After they have used up the bulk of their hydrogen fuel, gravity sets in motion an irreversible process. The star's core collapses and the temperature rises so rapidly that a tremendous explosion ensues, during which the star can reach up to 100 million times its original brightness. The resultant release of energy scatters what is left of the star's mass several light years into space at velocities of thousand of kilometers per second. This is what is known as a supernova.

Following these dramatic events all that remains of the original star is a small, extremely dense residual body, which is composed almost entirely of neutrons. Neutron stars are only about a dozen kilometers in diameter, but their mass is of such high density that one cubic centimeter weighs millions of tons. This is the equivalent of a supertanker squeezed into the head of a pin! Stars like these no longer emit any visible light, but do emit rapid bursts or pulses of energy, which can be detected with radio telescopes. This is caused by their extremely rapid rotation, ranging from milliseconds to several times per second, and the emission of intense beams of radiation, similar to the beam of a rotating lighthouse. Such stars are aptly called pulsars. The residual core of the Crab Nebula, the remnant of the supernova observed in 1054, is a notable example.

The North America Nebula (NGC 7000). An immense hydrogen cloud illuminated by neighboring stars.

An ultra massive star may continue to collapse even beyond the neutron stage and become a truly unusual object, a black hole. The force of gravity in such objects is so strong that even light can no longer escape it. Although they are not directly detectable, there is good indirect evidence that black holes do in fact exist.

MESSIER, NGC, IC, WHAT DO THESE DESIGNATIONS INDICATE?

Early observers noted so many types of nebulae, diffuse, dark, emission, planetary, that it became necessary to catalog them in some manner. Charles Messier was a 17th century French amateur astronomer fascinated by comets. In order to not to confuse existing with new comets, Messier carefully listed all other diffuse astronomical objects he found. This is how the Messier Catalog began, a list of 103 diffuse objects preceded by the letter M: M 1 for the Crab Nebula, M 42 for the Orion Nebula, etc.

However it soon became necessary to develop a more comprehensive catalog as more and more diffuse objects were discovered. This led to establishment of the New General Catalog in 1888 by the astronomer Dreyer, listing 13 226 objects. These are all designated as NGC objects. The most recent effort along these lines is the Index Catalog (IC), which contains objects that for the most part are accessible only to large professional instruments.

Lights in the city

*H*istorically much astronomy was done in or near major cities. Only near the end of the 19th century in fact was the value of locating observatories on remote mountaintops fully appreciated. Due to the thinner, drier air and the absence of smoke, rain, and smog at such locations, telescopic performance far exceeded what could be done at lower elevations. It was also toward the end of the 19th century that cities widely adopted that new invention, electric lighting.

Sidewalk astronomy

Despite the progressive deterioration of city skies due to streetlights, billboards, and other sources, all greatly hampering our ability to see the stars, many amateurs persevere and continue to observe from urban settings. The terms commonly used to describe the quality of the sky when observing are "seeing" (atmospheric steadiness) and "transparency" or atmospheric clarity. There are also many amateur astronomy clubs and associations in most cities, that take an active role in promoting astronomy.

In the 1960s, John Dobson, now one of the most celebrated amateur astronomers in the world, together with a group of enthusiastic friends, began a group in San Francisco called the "Sidewalk Astronomers." Their goal was to make astronomy accessible to everyone by literally bringing it to the streets and public parks, and showing people everywhere the wonders of the heavens. That is how street astronomy was born.

Light pollution

Children growing up in the last two decades of the 20th century are probably the first generation in history to be raised in an environment in which the stars are totally absent from the night sky. Indeed, as night falls, the stars are usually the last thing modern city people notice instead of the first.

"GOOD" LIGHTING AND "BAD" LIGHTING . . .

For a number of years now cities have installed streetlights that are particularly aggravating for urban astronomers. The worst offenders are unshielded bulbs or globes, whose incandescent light sources are directly visible. This type of fixture shines light in every direction, including straight up, and emits so much glare that it actually defeats the intended purpose of illuminating the ground. This type of lighting also prevents dark adaptation and, by causing the pupils to close, greatly diminishes visual acuity at night. Shielded lighting is far better in this regard and
helps direct most of the light toward the street where it does the most good. Being more efficient, this type of lighting can also be of lower wattage. This is more economical and comfortable for the eyes and actually provides more security.

Excessive light pollution combined with high humidity can make the sky so opaque that no observing is possible.

Many people over 50, however, still remember being able to glimpse the stars and even the Milky Way from city skies, and at least recognizing the principal stars and constellations. Public and commercial lighting might well highlight our tall buildings and monuments, and even instill a measure of pride in us, but, sadly, the stars have all but vanished from the sky because of that. Urban expansion has progressed at such a rapid pace the last few decades that virtually all cities today are enveloped by brilliant domes of artificial light which make any kind of astronomical observing extremely challenging. Astronomers call this problem "light pollution." Light pollution is even more prevalent under adverse atmospheric conditions, such as high humidity, dust, particulates, and smog. Light is not only reflected upward more directly under these conditions, but also becomes diffused throughout the city's dome.

Luminosity and magnitude

Despite this somewhat pessimistic scenario, many astronomical objects are actually visible from cities, as long as you take advantage of the most opportune time to observe them. Ancient astronomers grouped the naked-eye stars into six "sizes" or magnitudes based on

ANTI-LIGHT POLLUTION ASSOCIATIONS

The United States was the first country to draw attention to the touchy problem of light pollution. The International Dark Sky Association (IDA) has had considerable success with light abatement efforts in cities like Tucson, Arizona, where shielded street light fixtures have been installed with low pressure sodium emission which is easily filtered out. This type of enforced light restriction has greatly helped the work of nearby astronomical observatories. Similar organizations are being formed in many other countries. In France, the National Committee for the Protection of the Night Sky is seeking sensible public policy solutions to limit and prevent light pollution for general benefit.

their apparent brightness: these ranged from 1 for the brightest stars all the way to 6 for those barely visible to the eye. These general designations have been retained through the centuries, however, 0 magnitude has been used for objects brighter than 1st magnitude, with the brightest reaching into negative numbers. For example, Sirius, the brightest star in the sky is −1.6, while the Sun is −27. Whole magnitudes are actually expressed in geometric intervals of approximately 2.5, where a star of 2nd magnitude is actually 2.5 times fainter than a 1st magnitude star, and so on.

Weather conditions

What will the weather be like tomorrow? This is a question we ask ourselves almost every day, since weather affects everyone and often influences our daily actions. Usually, however, only astronomers are really interested in the weather for the coming night.

What constitutes a really good night for astronomy?

An in-depth discussion of meteorology is clearly beyond the scope of this book. It is important, however, for you to familiarize yourself with some basic weather indicators to help you decide in advance if it is likely to be a good night for observing. Sky conditions can change significantly after sunset due to such factors as atmospheric inversions, sharp drops in temperature, or changes in wind direction. Smog, dust, and hydrocarbons in the air can also impact observing conditions. Such factors vary seasonally and can really affect the transparency of the sky or create localized mini-climates.

Unless you live in a city with favorable weather all year round, you may have to pick and choose "good" nights for observing. And what constitute really good conditions for astronomy? Ideally, of course, you want a cloudless night, with no fog or haze, and no atmospheric turbulence. Nights like that are going to be rare. However, nights with light fog and no wind can be ideal for lunar and planetary observing. The stars might be almost invisible under such conditions, but seeing can be very steady, allowing you to use high magnification for the Moon and planets. Predicting good observing conditions in advance can be tricky, since even a totally clear day does not guarantee a crystal clear night and no turbulence.

Atmospheric highs and lows

Few factors affect regional weather patterns more than areas of high and low atmospheric pressure. In the Northen Hemisphere, high-pressure cells usually rotate in a clockwise direction and in summer often separate cool, northen air masses from their warmer, more humid southern or tropical counterparts. The

Will it be clear tonight? Careful attention to the sky at twilight is the urban astronomer's first concern.

Does this beautiful sunset promise clear skies later? No, unfortunately these strato-cumulus clouds are solidly in place and unlikely to move on any time soon.

boundary between the two air masses constitutes a weather "front." If the lighter, warmer air mass manages to push the colder air out, a warm front is generated. But if the cold air mass is stronger than the warm air and passes under it, a cold front is produced. The interaction between warm and cold fronts can generate instability and greatly affect atmospheric pressure. In general, high-pressure cells indicate fair weather, while an approaching low signals the arrival of poor weather. A barometer can be helpful in monitoring changes in atmospheric pressure.

Hot, humid air rising above a cold front forms high cirrostratus clouds.

A light fog indicates absence of wind, conditions that can provide excellent views of the Moon and planets.

Interpreting the barometer

In 1643, Torricelli, a student of Galileo, discovered the principle of the barometer. He noticed that the height of a column of mercury in a glass tube, with one end sealed and its open end inverted over a container of the same liquid, rose and descended in response to changes in atmospheric pressure. These variations turned out to be good indicators of changes in weather. The exact atmospheric pressure at any given time is less important than the overall trend.

Under stable, high pressure (e.g. above 30 inches of mercury or 1025 hPA or hectopascals), fair weather is likely to predominate. A progressive decline of 2–3 hPA over say three hours, indicates a change in weather, and a drop of more than 5 hPA per hour augurs only bad news. For example, in France on December 26, 1999, the air pressure dropped to and held at 950 hPA for several hours, resulting in hurricane force winds in excess of 200 km/hr, with disastrous consequences.

In short, by paying attention to the barometer, cloud patterns, wind directions and seasonal effects, you can predict with surprising accuracy what sky conditions to expect for the coming evening.

Spring Skies

It is often said that springtime brings fair weather. However, this is also a time of rapid weather changes and frequent showers. On clear days, temperature inversions at twilight can induce rapid atmospheric cooling, causing puffy cumulus clouds to dissipate like magic just as darkness sets in. Sometimes as well spring showers have the effect of "cleansing" the air, by completely removing wind-blown dust and particulates.

High atmospheric pressure in winter, with cold clear skies showing high altitude jet trails. The night sky will be overcast

Cumulu-nimbus clouds during late summer afternoon, a risk of sudden downpour

Big, puffy cumulus clouds are fair weather clouds and are likely to dissipate as night sets in

This can result in crystal clear skies right down to the horizon. Although telescopic images will be sparkling under those conditions, they will usually be quite turbulent as well. Once low-hanging strato-cumulus clouds make their appearance, a change in weather is in store and rain is almost inevitable. You will just have to be patient on days like that.

Summer skies

In the summertime, an absence of clouds and blue skies during the day does not necessarily guarantee clear nights. Paradoxically as well, city observing conditions are often least favorable at this time of year. That is because sunlight heats up pavements and concrete during the day, which is radiated back again at night, causing tremendous air turbulence. This completely degrades telescopic images, similar to distant objects shimmering above overheated highways. Often too, high air pressure under these conditions traps humidity and prevents cloud formation. The result is sunny weather, but no wind and a steadily stagnating atmosphere, which increases air pollution and decreases transparency.

This is when a big shower or thunderstorm can really help "clean" things up. The caveat is that this can also lead to thick cloud formations, which persist throughout the following night. Summer air pollution makes astronomical objects appear washed out, particularly if accompanied by turbulent seeing as well. The long summer days also go hand in hand with drawn-out twilight (and correspondingly shorter nights), particularly at higher latitudes. Fortunately, while summer is not usually the best season for urban astronomers, it is vacation time for most people and you can escape the city for better locations elsewhere.

10/03/2019

Item(s) Checked Out

TITLE   Photo-guide to the
BARCODE   33029038433124
DUE DATE   **10-24-19**

TITLE   Urban astronomy /
BARCODE   33029054069174
DUE DATE   **10-24-19**

Total Items This Session: 2

Terminal #   206

Autumn skies

As days start getting shorter again in the fall, things also start improving for astronomy. While the beginning of fall can be a lot like the "dog days" of summer, late autumn can extend into "Indian summer" well into October. After the autumnal equinox, changing winds and increased precipitation often yield clear, pleasant nights, well suited for observing. Days with puffy, white cumulus clouds and deep blue skies are especially good, since good, clear, dark nights will usually follow. However, the appearance of strands of high cirrocumulus clouds is not good for stargazing.

Winter skies

Winter skies offer a very rich assortment of interesting astronomical objects. Not only are many of them exceptionally beautiful, but the cold weather also provides some of the best sky conditions. If a cold front and high pressure set in, clear weather can persist for long periods of time. It might turn very cold, but also clear and cloudless. Moreover since the days are shorter in winter and darkness sets in sooner, you can enjoy considerably longer observing sessions. Under dry and cold winter conditions, the transparency of the sky will be at its best, allowing you to observed some of the brighter nebulae even from the city. You will soon discover the pleasures of winter observing, even though you will have to dress accordingly if you want to stay warm and comfortable.

Observing clouds

Our short course in meteorology has shown that urban astronomy is very much an exercise in patience, and that having to wait for those great observing nights is part of the game. Try to remember a few rules of thumb. A very

WHAT IS THE "GREENHOUSE" EFFECT?

The Earth is heated by infrared rays from the Sun. Do not confuse infrared with ultraviolet rays, which are also invisible and which are primarily responsible for summer tans. Infrared light rays penetrate the atmosphere and are reflected back by the Earth's surface. This affects a slight shift in wavelength, causing them to be reabsorbed by clouds, water vapor, and air pollution. As a result these rays do not bounce back into space, and the atmosphere traps their heat energy, similar to what happens in a green house.

The progressive rise in the Earth's average temperature is of concern to scientists and others. Air pollution can aggravate the situation and create "microclimates" over cities, for example, with abnormally high seasonal temperatures. Automobile emissions are a major contributor to air pollution. Such emissions can magnify the green house effect and cause health problems for many. That is why many states are imposing emission control measures and asking people to drive less or use public transportation.

clear horizon often indicates the arrival of rain. The sky may clear after a good rain and, although the transparency may be good, the air is often turbulent under those conditions. If distant sounds, like church bells or train whistles can be heard clearly, that means the air is dense and probably humid, often not a sign of good weather.

Equipment for city observing

Binoculars, an incomparable tool

Because they are usually small and compact, binoculars may not appear ideal for astronomy. Do not be fooled, however, these modest instruments can be your most valued stargazing accessory.

Their most important qualities

Binoculars are definitely not "simple" instruments, nor merely a beginner's entry to a "real" telescope. On the contrary, good binoculars may turn out to be your most used piece of equipment. In fact, many amateurs have actually done research-level work with binoculars, particularly in comet hunting and variable star observing. These compact instruments offer a mix of features most others do not: reasonable cost, immediate access and portability, wide field of view, excellent optical definition, comfort, and ease of use. Their most important feature, however, is the opportunity to use both eyes simultaneously when observing. This reduces eyestrain and lets you see far fainter objects than otherwise possible.

Their main drawbacks

Binoculars have two major shortcomings. In the dark and without reference to any ground-level objects, your arms tire quickly when observing with binoculars. This causes them to shake and star images to jiggle around. The only solution to this is to mount your binoculars on a camera tripod. That makes them less mobile of course and harder to use overhead near the zenith. Another shortcoming is that the magnification of binoculars is usually fixed between 7 and 15×, which limits their usefulness for observing details on the Moon or planets.

A mid-range telescope

As with any telescope, the diameter (aperture) of the main objective lenses is what really determines the performance level of binoculars. It is actually the **surface area** of the objective lens (or mirror) that determines its light gathering power. For example, binoculars with 50-mm diameter objectives, gather about 50 times more light than the naked eye. To gather 50 times more light than that you will need a telescope of 355-mm (14-inch) aperture. However an instrument that size can weigh 100 Kg (200 lbs) and cost as much as a small car. Clearly, therefore, our 50-mm binoculars fall about mid-point in performance between the naked eye and a large amateur telescope. In addition, because of their very wide field of view, many open clusters, comets, and some galaxies, often appear more spectacular in binoculars than in a telescope. Couple that with the stereoscopic effects that binoculars can provide, and you add an element of magic to your observing.

Methods and numbers

Binoculars are really two small telescopes mounted side by side. By using prisms in the optical path, they provide properly oriented images. In addition to opera glasses, with simple Galilean optics, there are two types of prismatic binoculars, those with **roof** prisms and those with **Porro** prisms. The former are more compact, but also harder to manufacture and hence more expensive. Porro prism

Optical path of roof prism binoculars

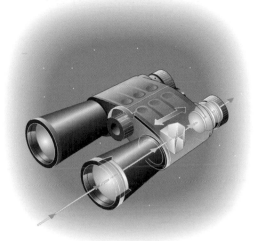

Binoculars with roof (left) and Porro prisms (right); the two most commonly available today.

Optical path of Porro prism binoculars

binoculars are widely available. Although neither type is really superior, in general roof prism binoculars tend to be high-end items.

○ **Performance**. Two numbers separated by an × are usually engraved on the binocular body. The first number refers to the instrument's magnification or power; the second to the diameter (in millimeters) of its objective lenses. For instance: 8×50, indicates binoculars magnifying 8 times and having an aperture of 50 mm. Numbers alone, however, do not tell the whole tale when comparing the performance of say, 7×50, 12×80, or 9×63 binoculars.

○ **Visibility factor**. To provide a helpful index of comparison, the Canadian writer Roy Bishop proposed a simple calculation called the "visibility factor" (VF). Just multiply the two numbers engraved on your binoculars by each other, and use their product as an index of night performance. For example, an 8×50 (VF400) is more powerful than a 7×50 (VF350), and a 12×80 (VF960) is almost twice as powerful as a 10×50 (VF500).

○ **Actual field of view**. Another important performance indicator for binoculars is their actual field of view. This number is usually also indicated in degrees: 6.5°, 7.5°, etc. Alternately, it may be shown as the width of the field covered in meters from a distance of 1000 m. For example: 131 m at 1000 m indicates the instrument covers a width of 131 m from a distance of 1000 m. To convert that to degrees in the above case, simply divide 131 m by 17.5, which rounds out to 7.5°.

- **Apparent field of view**. This is the field of view we see subjectively. It is calculated by multiplying the actual field of view by the instrument's magnification. Binoculars with a magnification of 10× and an actual field of view of 6.5° has an apparent field of view of 65°. This value is actually very important, because a large apparent field of view does much for observational comfort.

The exit pupil: a controversial number

The exit pupil is a small, bright circle of light visible in any pair of binoculars held at arm's length. This is really the effective aperture through which the light is funneled. Its size is easily calculated by dividing the objective aperture by the magnification. For example with 8×56 binoculars, the diameter of the exit pupil is 7 mm. This number is important since the larger the exit pupil; the brighter the image appears in a given instrument. Binoculars intended for night use should typically provide a 4–7-mm exit pupil. A typical pair of marine 7×50 binoculars will provide an exit pupil of 7.1 mm. A large exit pupil is also handy when you observe from an unstable of moving platform, such as a boat.

It is often said that only young observers can take full advantage of a 7-mm exit pupil, since the pupils of their eyes actually dilate that wide when fully dark adapted. After the age of 50, however, the human pupil typically opens no wider than 5 mm, blocking out much of the light transmitted through a 7-mm exit pupil. This contention is subject to considerable dispute, however.

You can go to an optician and check this out yourself by comparing the performance of a 7×50 and a 7×25 pair of binoculars in full daylight. They have exit pupils of 7 and 3.5 mm, respectively, but your pupils will shut down to 1.5 or 2 mm in bright light and act as

A cut above, these 22×60 Takahashi binoculars with fluorite lenses provide exceptional image quality.

equalizers. In theory, therefore, you should perceive no difference between these binoculars and yet which of the two provides the brighter image?

- **Eye relief**. Eye relief is the distance your eye must be held from the eyepiece in order to see the entire field of view. This distance is measured in millimeters and is a very important value to know, particularly if you wear glasses. Larger eye relief also makes for more comfortable viewing.

Are there ideal binoculars?

Binoculars intended mainly for astronomy will inevitably represent a compromise. To observe faint deep-sky objects, for example, you want low power and a wide field of view. To see more detail in any object, however, you really need some magnification. A wide field of view and a magnification of seven

Benefits of military technology

For those who desire the very best, Canon offers binoculars with image stabilizers. You simply focus, push a button, and the image is instantly rock solid! These instruments are equipped with gyroscopic senors, which detect any vibrations imparted by your hands. Electronic corrective signals are then transmitted to movable prisms in the light path, which deflect the light rays in real time to compensate for any vibrations. These binoculars also feature low-dispersion ED glass lens elements, which, like professional telephoto lenses, deliver superb, high contrast images right to the edge of the field of view.

These 15×50 Canon binoculars with image stabilizers are quickly becoming favorites with many observers.

may provide you with great views of some portions of the sky, but you might miss out on smaller, fainter objects. With 10 to 15× binoculars, you will see much detail in star clusters and bright nebulae, but your field of view will definitely be more restricted. From experience, we suggest wide angle, 10×50 binoculars for general astronomical use or the new 15×50 Canon IS. For astronomical use avoid zoom type binoculars completely, since their many additional lens elements can lower contrast and introduce internal reflections.

How much should you pay?

Binoculars are available in all price ranges, and often the best alongside the worst. Obviously, instruments combining excellent optical quality, a wide field of view, and good eye relief, will also be quite expensive. A good, mid-priced pair will start at around $100–150, while a top of the line model can cost $500–1000. But do not panic! You will only have to buy a good 10X50 pair of binoculars once. There are plenty of choice brands available: Leica, Zeiss, Fuji, Nikon, Bushnell,

Bausch & Lomb, Pentax, and more, and many excellent models in the $100–250 price range. Our advice? Try out as many as you like before you buy, since your final selection will be a lifetime investment.

A word about giant binoculars

Some binoculars can be downright intimidating. Anything above the 50–60-mm aperture range falls into the giant or jumbo class. Large aperture instruments like this were originally devised for surveillance and military use. However, binoculars like the 11×80 Vixen and other brands offering 20×100 and even 25×150(!) size instruments (Fujinon), can cost as much as a small car, with the weight to match, and provide stunning views of the heavens. Apart from costing as much as a really good telescope, these instruments also have to be properly mounted for effective use, since they are too heavy to be hand held and are really quite cumbersome to move or transport. You should only consider binoculars of that size if you are a serious comet hunter, for example, but then you might as well get a more suitable telescope instead.

Buying a telescope

You have become sufficiently familiar with the sky and are now ready for a real telescope. It is best not to go overboard on this just yet and keep in mind that the best telescope for you at this stage is not necessarily the biggest, most powerful or expensive instrument.

The function of a telescope

Contrary to what many might think an astronomical telescope is not a "magnifying machine," but really serves two main functions. First, it captures and amplifies the light of often-faint astronomical objects, so that they appear bright enough to be observed with the eye. Second, a telescope provides the "resolving power" needed to discern fine detail on the Moon and planets, for example. Both of these characteristics depend on the aperture or diameter of a telescope, rather than its magnifying power. In fact, a telescope's aperture limits both the magnification you can usefully employ as well as how faint an object you can see with it.

Various types of telescopes

There are basically three types of telescope:

● **Refractors.** These rely exclusively on lenses to collect and transmit light. Refractors were the first types of telescopes built, and in many ways the simplest in design as well. They usually involve a doublet of lenses at the front of the optical assembly and a smaller ocular or eyepiece at the other end, with nothing else in between.

● **Newtonian reflectors,** or simply reflectors. These use two mirrors to collect and transmit light. A large, concave mirror at the bottom of the optical tube collects light and reflects it back up to a smaller secondary mirror. The latter, in turn, directs it toward the eyepiece mounted on the side of the tube where the observer places his or her eye.

● **Catadioptic or compound telescopes**. These are generally of two types: **Schmidt-Cassegrain** and **Maksutov**, named after their respective designers. They are compact instruments, with large, concave primary mirrors at the back of the tube assembly and a central perforation to transmit light reflected by a smaller secondary mirror at the front end. Both designs also have special corrector plates or lenses at the front, which bend or refract incoming light rays before they reach the primary mirror. The main advantage of these types of instruments is their compact design. Since light rays are reflected back and forth within the optical tube, it becomes

---

LET'S TALK "MAGNIFICATION"

No matter what type of telescope you have, the highest effective magnification you can use should not exceed 2–2.4 times the diameter of its objective lens or mirror as measured in millimeters (or 50–60 times its aperture in inches). For example, a 100-mm (4-inch) refractor can comfortably handle 240×, depending on visual and atmospheric transparency of course. Anything higher than that will be "empty" magnification, producing fuzzy, low-contrast images. As a good rule of thumb, for the most part you will probably use a maximum magnification similar to that of the instrument's aperture in millimeters.

To calculate the magnification, simply divide the focal length of the telescope by the focal length of the eyepiece you are using. For example, using a 5-mm focal length eyepiece on a 900-mm focal length telescope gives 900/5 = 180×. Most telescopes come equipped with three fixed-focus eyepieces, providing a progressive range of magnification (see page 44).

possible to physically compact a long focus telescope into a tube four times shorter than a refractor or reflector of similar focal length.

It is important to remember that all astronomical telescopes invert the images they produce. Apart from being a little confusing, this is of no real consequence since "up" and "down" has little meaning in outer space. However, just in case, image erectors are available for refractors and Schmidt-Cassegrain telescopes, more commonly known as SCTs.

Aperture, focal length, and focal ratio

Two main parameters dictate the power or performance of a telescope, the diameter or **aperture** of its objective lens or primary mirror, and its **focal length**. Both values are typically expressed in millimeters for amateur class instruments, although the aperture is often also stated in inches, particularly in the USA. We will deal primarily with instruments in the 50–150-mm aperture range (2–6 inches), since in the city anything of 200-mm (8-inch) aperture or larger is really not justified.

A telescope's focal length is the distance behind the objective lens or mirror at which it forms an image of any object located at infinity. The focal length of the eyepiece, which is used to examine that image as with a magnifying glass, is also expressed in millimeters. This will typically range from 4 to 40 mm. If we divide the focal length (F) of any telescope by the diameter (D) of its objective (F/D), we obtain its focal ratio or f/r. The f/r affects the photographic "speed" of a lens or telescope in terms of how quickly film will record an image. Visually, a longer f/r is better in the city than a shorter one, since the background sky glow due to light pollution is not as apparent).

Short focal ratio instruments (e.g. f/4 or f/6) are physically shorter than longer focal ratio (e.g. f/10 to f/15) telescopes of the same

Project the image of a distant scene on your wall with a magnifying glass. The focal length of the magnifying glass is the distance between it and the wall

aperture. They are generally preferred for wide field, deep-sky observing, while the longer instruments are well suited for lunar and planetary work. Most commercial suppliers will indicate the above three parameters (aperture, focal length, and f/r) very clearly which greatly helps comparative shopping. The main question for you, however, will hinge on what instrument fits your particular observing circumstances and interests, as well as your purse.

FIVE GOLDEN RULES FOR BEGINNERS

1. Do not rush out to buy a telescope from just anywhere.
2. Do not be too concerned about size. A 100–120 mm (4-6-inch) telescope is an excellent compromise for use in the city.
3. Do not be fooled by the importance of "power" or magnification; those are totally secondary considerations.
4. Join a local astronomy club and get a chance to observe with a variety of telescopes before selecting your own.
5. It is generally best to buy from a reputable dealer who specializes in astronomical instruments and accessories, and who can help you in case of repair or service needs. Many dealers are also mail order and on-line suppliers (see page 105).

A worry-free telescope

Instead of binoculars, you have opted for a telescope right away. Before buying one, however, be sure to consider fully your needs and circumstances so that you will choose an instrument most appropriate for you. The choices and selections available are many and can be quite confusing unless you do your homework beforehand.

A good telescope for the city

As a city dweller, any telescope you get is likely to be used mainly from your backyard, driveway or balcony. Before you buy, therefore, apply some basic common sense to help you decide what size of instrument and performance level you want:

- What is the best instrument in my budget and price range?
- What is the best instrument for my location and available observing space?
- Aperture for aperture, which type of telescope is optically superior?
- Is an altazimuth mount adequate for my needs or is an equatorially mounted telescope essential?

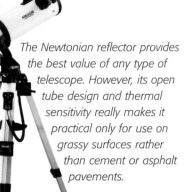

The Newtonian reflector provides the best value of any type of telescope. However, its open tube design and thermal sensitivity really makes it practical only for use on grassy surfaces rather than cement or asphalt pavements.

Unfortunately, in asking these questions you are inevitably faced with a compromise situation. No telescope will meet every situation, nor are you likely to find one that is optically perfect and ideally mounted, and also comes at a bargain price. Realistically, your "ideal" telescope should include the following essential features:

- Generally convenient and easy to use.
- Readily transportable from one location to another.
- User friendly and simple to operate.
- Equipped with a good mount that can be set up quickly and without hassle.
- The best optical and mechanical quality you can afford.

So what telescope is best for me?

Experience has shown over and over again that people will more readily use even just a "modest" instrument if it is easy to set up and operate. Please note that "modest" in this context does not imply a compromise in quality, rather it refers to modest size and power.

- Most people agree that you get the best value for your money with reflecting telescopes like **Newtonians** or better still **Dobsonians.** These designs provide the best return for your money in terms of aperture (and hence light-gathering power and resolution).
  However using a telescope like this from an apartment can be a challenge, since they tend

to have long, open-ended tubes, which are subject to internal air currents. In addition, the optics may require some time to adjust to often sharp temperature differentials between indoor and the outdoor air. Schmidt-Cassegrain telescopes are considerably less prone to these types of problems. Finally, Newtonians may require periodic collimating since their optics can get knocked out of alignment. This is something many amateurs are reluctant to do because it requires a fair amount of patience and skill. In short, unless you have access to a garden or backyard, we would not recommend a Newtonian as the telescope of choice for city observers.

◉ For all-round performance, value, and convenience, the **Schmidt-Cassegrain** or **Maksutov** designs are probably the best telescopes for urban observers. Thanks to their compact optical design, they take up a minimum of space. The optical tube assembly of a typical 20 cm (8-inch) f/10 SCT is only about 40 cm (16 inches) long, for example, while the tube assembly of a Newtonian of similar aperture and focal length would be closer to 2 m or over 6 ft in length. Although SCTs also require some time to thermally equilibrate before use, such times are shorter than for open-ended reflectors, nor are tube currents generally a problem.

On the downside, catadioptric telescopes require a rather larger secondary mirror. This constitutes a sizeable obstruction in the optical path and tends to lower image contrast somewhat, relative to other telescope designs. This drawback is largely overridden, however, by compact size and portability of these telescopes, making them very popular among amateur astronomers generally.

◉ From a purely optical perspective, **apochromatic refractors** ("apos," containing fluorite and other special glass elements) are by far superior to any other design aperture per aperture. Offering superb resolution and image

The Schmidt-Cassegrain telescope, plenty of performance in a compact package. A very popular optical design.

contrast, these instruments are without parallel on several levels. They also reach thermal equilibrium very quickly and so can be put to use almost as soon as they are set up.

A refractor can be a little cumbersome to use compared to other telescope types, since its optical path goes straight through the tube. However, collimation is almost never a problem since the lenses are properly and permanently aligned at the factory. Their major drawback? Price. It can be "astronomical" in comparison to other types on an aperture per aperture basis.

Turning next to classical achromatic refractors, they are considerably more affordable than "apos", but do suffer from residual chromatic aberrations. This tends to give images a slightly colored fringe or halo, particularly with very bright astronomical objects.

Depending on your budget and enthusiasm, either a 70–100-mm (3–4-inch) classic refractor or apochromat can be an excellent choice for the city bound observer. In conclusion, we recommend refractors or small catadiotric telescopes over Newtonian-style reflecting telescopes for urban settings; primarily because the latter suffer from tube currents and thermal fluctuations.

Equatorial or altazimuth mount?

No matter how good your telescope, ultimately it is only as good as the mount on which it is used. When it comes to mountings, mechanical stability is as important as optical quality is for the telescope, since nothing is more annoying than a wobbly mount that shakes in the slightest breeze. There are essentially two types of telescope mountings, altazimuth and equatorial. Which should you choose?

Selecting the right mount

The simplest commercial telescopes are usually supplied with altazimuth mounts. The operating principles of this type of mount are quite straightforward; there are two axes, one for vertical, and the other for horizontal movement. The telescope is typically held by a yolk-type arrangement which can be rotated 360° and which lets you point the tube freely from horizon to the zenith overhead.

Remember, however, that the Earth rotates

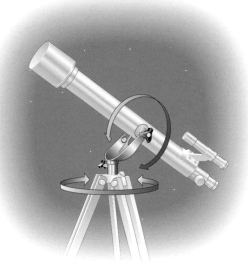

Operating an altazimuth mount is relatively simple and quite intuitive, which explains why it was considered ideal for beginners. Despite its shortcomings, this type of mount is useful for small telescopes used from cramped urban locations.

steadily about its axis. This results in an apparent motion of all objects in the sky from east to west. Too small to be noticeable in binoculars, such motion becomes quite apparent, however, with a telescope which magnifies and hence amplifies the effect. At $100\times$ for instance, this motion appears 100 faster than to the naked eye. Consequently, to keep an object centered in the field of view with an altazimuth-mounted telescope, you must constantly compensate for this through incremental adjustments in both horizontal and vertical directions. This becomes even more aggravating with higher magnification.

This situation can be improved somewhat by inclining the mount's vertical axis as closely as possible to the latitude of your location, relative to the sun, and by keeping it level with that angle. This way you can track astronomical objects more easily by rotating the telescope about the inclined axis in an east–west direction. This is basically how an equatorial mount works (and is in fact what you created with this realigned altazimuth).

Finding any object in the sky

Equatorial mounts are designed with two axes, a polar (or hour) axis and a declination axis. The latter corresponds to the original horizontal axis of the altazimuth design. The two axes on most equatorial mounts are equipped with graduated setting circles, which allow you to locate any object in the sky by its

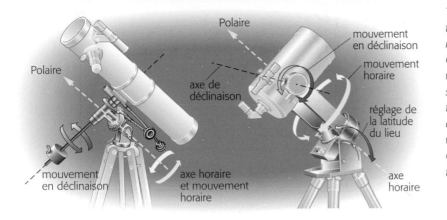

Polaire

mouvement
en déclinaison

mouvement
horaire

axe de
déclinaison

réglage de
la latitude
du lieu

Polaire

mouvement
en déclinaison

axe horaire
et mouvement
horaire

axe
horaire

*The most popular
types of equatorial
mounts are the
German Equatorial,
with a "T"-like
structure (left), and
the fork mount
commonly provided
with Schmidt-
Cassegrain
telescopes (right).*

two coordinates, right ascension and declination.

In practical terms, what you do is turn on the telescope drive and point the mount's polar axis as close as possible to Polaris the North Star. Located near the north celestial pole, this is the spot in the sky where the northern end of Earth's axis is pointing. Most quality equatorial mounts today are supplied with a small alignment scope fixed in the bore of the polar axis and equipped with an illuminated reticle to center the mount precisely on the celestial pole. This is important since Polaris itself is not located exactly at the celestial pole but is offset from it by about 48 arc minutes. A polar alignment scope compensates for that and points the mount at the true pole.

Unfortunately, to take full advantage of this feature, you must be able to see Polaris from your location, something not always possible from a balcony or other restricted observing space. Often as well, beginners are reluctant to properly align their telescope mounts because they think this is technically too difficult to do. This is really not the case, however. A properly aligned equatorial mount with clock drive will greatly add to your viewing pleasure (and comfort) by tracking objects for extended

periods without needing readjustments. A set up like this is also essential if you are contemplating any kind of astro-photography or imaging.

Altazimuth-mounted telescopes

Incredible as it may seem, most leading telescope manufacturers, including Celestron, Meade, Losmandy, Astrophysics, and Vixen, now offer altazimuth-mounted telescopes with all the state-of-the-art features listed above. Celestron's "Nexstar" series of telescopes even includes a guided "tour" function with its mounts, complete with verbal descriptions of your favorite section of the sky! This sophisticated technology is an extension of the automated guidance systems developed for modern, altazimuth-mounted telescopes at leading professional observatories. Gigantic telescopes, like the VLT (*Very Large Telescope*), the Keck and Gemini giants, are simply too large to be equatorially mounted. Thanks to computers and modern guidance technology, the smaller and less costly altazimuth style of mount has now become the norm for professional observatories as well . . . and amateurs too have reaped the benefits.

Let's talk budget . . .

*A*s with many other things in life, "you get what you pay for" is certainly true when it comes to astronomical equipment. Remember a good telescope may be more expensive initially, but you risk becoming disenchanted with a mediocre performer and give up observing altogether because of that. That may "cost" you even more in the end . . .

Your budget is really small . . .

"I only have a few hundred dollars to spend but I really want that telescope.."

◉ We suggest you buy yourself a good pair of 8×50 or 10×50 binoculars instead. Both Nikon and Pentax offer 10×50 models, the CF Action and PCF, respectively. The latter is definitely a best buy in this price range. There are many quality performers available in the 8×56, 9×63, etc., categories supplied by Orion, Celestron, and other manufacturers and distributors. Be sure to try out as many types and styles you can first hand, checking them out for comfort, contrast, sharpness, eye relief, and field of view.

◉ If you cannot sleep at night because you are really interested in lunar and planetary observing, we suggest getting a small, good-quality achromatic refractor instead. Many are available in the 60, 70, 80, and even 90-mm (2.6–3.5-inch) aperture range from suppliers like Celestron, Meade, Orion, and others, costing between $50 and $250.

Your budget is small . . .

" I want a quality instrument but can't spend more than $400–500".

◉ You have a lot more choice here. You might also consider a pair of 10×70, 16×70, or 12×80 giant binoculars in this price range. Again, there is a wide selection of models and brand names to chose from, including Celestron, Fuji, Canon, and Orion, as well as many distributor and commercial brand names.

◉ Several really good refractors are available in the 70–90-mm aperture range from Celestron, Meade, Orion, and other manufacturers. Several of these are equipped with equatorial

The Meade model DS 80 is good value for good quality.

TOO MUCH APERTURE

For urban astronomers we usually suggest compact telescopes not exceeding 150 mm (6 inches) in diameter. Anything larger than that only amplifies light pollution and exacerbates sky glow, which help to reduce image contrast.

mounts, solid tripods, and many other accessories. There are also a number of very compact 80–90-mm (f/5) refractors available. Although their achromatic objectives may exhibit some residual color, they are well suited for wide-field observing of deep-sky objects, including star clusters and bright nebulae.

Although altazimuth mounts have long been considered a mount for beginners, recent technological innovations have brought that design back to the forefront. Thanks to several developments in computer-driven location technologies like "GO-TO", the altazimuth mount is definitely making a comeback. What could be better than a computer-driven mount with rapid slewing motors that can automatically find any number of astronomical objects and point the telescope at them? All you have to do is enter the date and time on your mount's computer and select a couple of prominent stars to let the system orient itself. After that, the GO-TO operation will find any of several thousand or more objects in the sky whose coordinates are in its memory. For example, pick any of the 110 Messier objects or the 7800 NGC objects, and press GO-TO. Your drive will quickly slew the telescope to the appropriate coordinates and place the object directly in the field of view of your eyepiece. What planets are visible tonight? Your telescope's computer can tell you right away.

⊙ Meade, Celestron, and other manufacturers also offer altazimuth-mounted refractors like these with dual axis, motorized drives, and Autostar and NexStar GO-TO capacity. This makes it easy to automatically locate and observe thousands of celestial objects stored in the instruments' data banks.

⊙ Meade ETX 90 EC: This is an ideal first telescope for many and comes equipped with a computer driven, dual-axis altazimuth mount. (It is a Maksutov design), with excellent optics

and, at f/14, especially well suited for lunar and planetary observing. Despite its rather plastic-like finish, this telescope tracks well and can also be equipped at additional cost with the Autostar guidance system. The ETX has legions of fans worldwide and is featured on several Internet websites.

⊙ Takahashi FS 60: This small, 60 mm f/6 refractor is in a league of its own and performs like a much larger telescope. Have you ever observed with a telescope containing fluorite lenses? Fluorite is a crystal of CaF_2, calcium difluorite, and optics containing them provide almost completely color-free images. Moreover, the images are exceptionally sharp and contrasty and literally "snap" into focus in ways not possible with traditional achromatic lenses. One look through a fluorite lens will convince you

The ETX 90 is a very popular Maksutov telescope on both sides of the Atlantic ocean.

The Takahashi FS 60, a treat for the eyes but it comes with a price.

The venerable C-5 continues to adapt to new uses and technologies.

of their clear superiority. However, at about $800 for the tube assembly alone, the FS 60 is pricey and represents a significant investment.

Your budget is mid-range . . .

"I can invest about a thousand dollars, what can I get for that?"

◎ Once again, consider getting a really good pair of binoculars. Despite having a somewhat narrow field of view the Takahashi 22×60 with fluorite lens elements is about tops in its class. These are in effect two 60-mm refractors (similar to the FS 60 above) but mounted side by side. Because of their size and 22× magnification they must be used with a tripod to remain steady. You might also consider getting the Canon 15×50 IS binoculars with image stabilizers.

◎ The venerable Celestron C-5, Schmidt-Cassegrain telescope is a very good choice in this category. Its 125 mm (5 inch), f/10 optics are good enough for high magnification work (up to 300×) and make it a very versatile instrument. It can be bought as a spotting scope tube assembly only, or fully equipped with equatorial mount, tripod, drive, and drive corrector. This instrument's quality optics and reasonable price rated it a "best buy" by BVD, the well-known

American birdwatcher association. Depending on model and accessories, this telescope is available at $700–1000.

◎ The Meade ETX 125 mm or the Celestron Nextar 5, are instruments of comparable quality and performance. The Nextar is basically the C-5 optical tube assembly. However, it has an unusual single arm fork mount with a GO-TO drive system. Thanks to its relatively faster f/10 configuration, it is somewhat more versatile than the ETX. Like the ETX 90, the latter has a f/15 focal ratio, which really limits it primarily to lunar and planetary work. The Celestron 5 can be used with lower power, wider field eyepieces, essential really for most faint deep-sky objects. The EXT 125 sells for around $900 and the Nextar 5 with tripod for about $300 more.

◎ Televue Pronto and Ranger: designed and conceived by an optical engineer involved in the Apollo lunar mission, Televue products have a real "space age" quality about them. The Pronto and Ranger refractors are optical and mechanical jewels of the highest quality. Both feature the same high-quality, 70 mm, f/6.8 objectives with ED (low dispersion glass). The main difference between the two models is in

The Meade ETX 125, is similar to the ETX 90 but with twice the light gathering power.

The Televue Pronto, exceptionally sturdy construction and with unmatched optical quality.

their finish and focusing mechanisms (the Ranger has a simpler helical focuser). In addition to their refractors, Televue's range of eyepieces is also exceptionally good. In combination the two telescopes give the impression that you are observing with much larger telescopes than you are actually. That is part of the reason for their universal acclaim. The Ranger typically sells for about $700 and the Pronto for around $1200.

○ A number of refractors of Chinese manufacture are also available in this price range from Orion and other distributors, including the AstroView (f/5) and medium fast (f/8), 120-mm (4.7-inch) aperture achromats. These are solidly mounted and perform well both visually and photographically and come with appropriate accessories.

You have an unlimited budget . . .

"I really want the best and am prepared to pay for it. . .".
○ For the urban observer there is little point in buying a really large aperture telescope; go

for top quality instead. Takahashi, Televue, Astrophysics, Orion, and many others, offer apochromatic or near apochromatic refractors with fluorite and ED glass elements of exceptional quality. Most of these are in the 100-mm (4-inch) aperture range, with fast (f/5.5) to medium fast (f/6–8) optics, but well suited for all types of observing from lunar and planetary to deep sky. The contrasty, color-free and extremely sharp images provided by these telescopes are the absolute best. Be prepared to pay anywhere from $2500 to $5000 and more for these instruments, however, often for the tube assembly only. Accessories do not come cheap either.

The Takahashi FS 102, provides outstanding images second to none.

Eyepieces

An eyepiece or ocular functions like a magnifying glass at the focal plane of the telescope. Its purpose is to enlarge the image and render it visible to the eye. The caliber of an eyepiece is just as important as that of the objective lens or mirror, since it can either enhance or degrade the overall optical quality of the final image.

The other "half" of your telescope

Most commercial telescopes are sold with one or two eyepieces. This is fine to get you started. Pretty soon, however, you will feel the need for additional oculars and a wider range of magnification. Purists contend that you must have at least four or five different eyepieces for this purpose. In reality though, you will probably end up using the three eyepieces below most of the time:

- A minimal or low power eyepiece, with a magnification equal to 0.2–0.4 the diameter (D) of the objective lens or mirror in millimeters.
- A medium power eyepiece, or one providing a magnification of 1–1.2 D (for optimal resolution).
- A high power eyepiece, one giving a magnification around 2 D.

As examples with a telescope of 90-mm (3.5-inch) aperture, a low-power eyepiece should magnify between 18 and 35×, a mid-range eyepiece between 90 and 110×, and for high-power eyepiece about 180×. With a telescope of this size, you would rarely use anything much higher than that.

Calculating the magnification

Just as with telescope objectives, focal length is a key characteristic of eyepieces. The focal length of eyepieces is expressed in millimeters and is usually engraved on the end of the barrel. You can calculate the working magnification (M) of any eyepiece by dividing

A star diagonal and an assortment of 3 or 5 quality eyepieces will meet most basic needs of the typical amateur astronomer.

its focal length (f) into the focal length (F) of the telescope objective lens or mirror, or M = F/f. For example, using a 20-mm focal length eyepiece with a telescope of 800-mm focal length gives a magnification of 40×. A 5-mm eyepiece on the same scope would give 160×.

Calculating the exit pupil

As indicated previously with binoculars (p. 32), various eyepiece/telescope combinations provide different size exit pupils depending on the magnification produced. The higher the effective magnification, the smaller the exit pupil and the dimmer the image appears. Conversely, the lower the magnification, the larger the exit pupil and the brighter and more comfortable the image will appear to the eye. Clearly then, the exit pupil helps define the working limits of your instrument.

An exit pupil larger than 7 mm cannot really be exploited fully by the average person, and one smaller than 0.5 mm will result in very dim, hard to focus images. It will also make you aware of "floaters" and other debris in the vitreous humor of your eye.

To calculate the exit pupil of any optical system, simply divide the diameter (D) of the telescope objective by the magnification a given eyepiece provides. Using the example above, if the diameter of the telescope objective is 100 mm and you are using a magnification (M) of 40×, the exit pupil is equal to (D/M) or 2.5-mm.

Calculating the actual field of view

Due to design and other parameters, eyepieces can have apparent fields of view ranging from about 45° to 80°. You can calculate the actual field of view of any eyepiece by simply dividing this value by the magnification. For example, a 15-mm focal length ocular with a 52° apparent field of view and a working magnification of 53×, provides 52°/53 = 0°58'52", or an actual field of view slightly under 1°.

You can also measure the actual field of view of any eyepiece/telescope combination directly, by determining how long it takes a star near the celestial equator to transition across the field of view. (Be sure to shut off the clock drive on your telescope mount.) By dividing the transit time (expressed in seconds) by 4, you can calculate the actual field of view in minutes of an arc.

Various types of eyepieces

A large assortment of eyepiece designs and brands is available today. Without going into a lot of detail, you should know that the classic "orthoscopic" and Plössl designs represent the best value for quality/performance/price for amateurs seeking quality.

Meade, Celestron, Vixen, Orion, Televues, and Takahashi, all feature a wide selection of orthoscopic and Plössl designs of good quality, available in a wide range of focal lengths from 4 mm to 50 mm. They are typically priced between $40 and $150. For the very exacting observer who can afford them, Meade offers its UWA (Ultra Wide Angle) series, and Nagler features the Panoptic and Radian eyepieces. These are exceptionally fine products and often literally transform the performance of a telescope. A point to note is that most standard eyepieces today have barrel diameters of either 31.75 mm or 50 mm (1.25 or 2 inches). Some entry-level telescopes are supplied with smaller eyepieces that are 24.5 mm (0.96 inch) in diameter.

THE BARLOW LENS A USEFUL ACCESSORY

If you are the happy owner of a short focal length and fast focal ratio (f/4-5) telescope, it may be difficult to attain magnification as high as you would like on occasion (even with very short focal length eyepieces). This is where a special lens called a Barlow comes in. This is a concave lens (with a negative focal length), that can double or triple the magnification of any eyepiece it is combined with. So, even if you have only three eyepieces in your collection, with a Barlow lens you will effectively have six or even nine eyepieces of different focal lengths.

The sky above our roof tops: the solar system

Asking for the Moon?

The sky is darkening, the city has gone quiet and the Moon is becoming visible amid the urban haze, out shined only by brilliant Jupiter. Quietly and oblivious to the rest of the world, you set up your telescope in a darkened corner and are ready to observe. At times like this and for you alone, the night is like magic.

The most observed object in the sky

Few astronomical objects are more familiar to people generally than the Moon. It is also among the objects most studied by amateurs; even experienced observers familiar with so many other astronomical objects cannot resist revisiting our closest neighbor in space from time to time. And, yet, how familiar with it are we really? As we shall see, the Moon offers us some truly spectacular vistas.

Our satellite is probably the Earth's younger sibling. Like it, the Moon is thought to have formed some 4.6 billion years ago through accretion of dust and residual material orbiting our planet in its formative stages. An alternate possibility is that the Moon formed after a wandering planet-size body collided with the young Earth, which then recaptured the resultant debris.

A dead world

Due to its relatively small size and mass, the evolving Moon was not able to hold on to the bulk of its original gasses and so could not retain a significant atmosphere. Without atmospheric pressure, there can be no liquid water or any other ingredient necessary to sustain life. This situation has probably persisted for several billion years. The Moon is not only a truly dead world, therefore, but also one without the benefit of an atmosphere to shield it. As a result, it has been subject to unabated meteoric bombardment for much of its existence, as clearly manifested by the more

THE FORMATION AND NATURE OF LUNAR "SEAS"

The Italian astronomer Riccioli drafted the first map of the Moon in 1651. At that time people referred to many lunar features by familiar mapping terms like oceans, seas, lakes, gulfs, etc., unaware that liquid or flowing water had not ever existed on the Moon, nor that most of the visible craters were not like volcanoes on Earth. Consequently, many of the historical names of prominent lunar features were established by analogy with terrestrial geology and have been retained to this day. In fact, it was not until the Apollo lunar missions in the 1979s that the real physical and chemical characteristics of the Moon were fully established, including the complete absence of water, trace elements of rare and ancient gasses trapped in lunar rock samples, extremely ancient traces of volcanic activity, etc.

than 300 000 craters on its visible side alone. Many of these impact craters and rings are 250 km and more in diameter. Many lunar mountains and peaks are proportionately higher on average than those on Earth, giving them sharp relief and rugged appearance. Likewise many lunar crevasses and canyons also assume proportions unheard of on Earth.

Twin sister planets?

The Earth–Moon combination is actually quite unusual in many ways and certainly unique in the solar system as a whole. Seen from nearby Venus or Mars, our home planet and its satellite look more like twin planets, whose maximum angle of separation does not exceed 1°. This must be quite a sight but one which most of us will never be able to see.

The Moon revolves around our home planet in about 29 and one half days. This is also the amount of time it takes the Moon to rotate once about its own axis, explaining why it always presents us with the same side or face and also why it has a side we can never see from Earth. This type of rotation is called synchronous rotation.

Because of these orbital characteristics (and depending on its position during a given lunation), the Moon appears to set or rise about 50 minutes later on successive nights, and never at exactly the same location in the sky. This delay is due to the Moon's orbital motion from west to east in its 29 and one half day revolution around Earth, and results in a progressive change in the angle of illumination from the Sun. We see this as a gradual change in the phases of the Moon, which are really nothing more than variations in our line of sight.

A 15-day tour of the Moon

To begin your tour, get yourself a calendar that includes information on the pivotal stage or "age" of the Moon each month. Each lunation begins with the New Moon date (often illustrated by a black disk). There is really nothing to see at that stage, since our satellite is in conjunction with the Sun at that point and too close to it to be visible.

About 48 hours after the New Moon, start looking for it with binoculars shortly after sunset and you should see a very thin crescent just above the western horizon. Things start getting really interesting about three days after the New Moon date. At that point, you can see a dimly lit lunar disk, sharply outlined by a brilliantly illuminated crescent. If we could look back at Earth from the Moon itself at that point, we would see a dazzlingly bright planet totally illuminated in its "full Earth" equivalent. This light is reflected back to the Moon, and lets us see its dark side dimly

illuminated by "Earthshine." This is best done with binoculars or with low magnification in a telescope. The Earth and the Moon exhibit phases in exactly reciprocal fashion. In other words, a hypothetical lunar inhabitant would see a quarter phase Earth at the same time we see a fourth quarter Moon.

During the following few days, pay particular attention to the slowly moving boundary between the illuminated and non-illuminated portions of the Moon. This sharp demarcation is called the *terminator* and indicates the changing angle of illumination as the Sun rises above the lunar horizon. This progressive shift in the position of the terminator casts the lunar surface in sharp relief.

Features not to miss

Refer to the maps on pages 52 and 53 to help you locate the most prominent features visible at various phases of the lunar cycle, and then correlate them with your calendar so that you can catch them at the most opportune time. About five days after the New Moon, the seas of Tranquility and Serenity are well placed for observation as well as the craters Piccolini, Theophilus, Cyrillus, and Catharina.

FACTS AND FIGURES ABOUT THE MOON

Diameter: 3476 km (2160 miles)
Age: 4.6 billion years
Mean distance from Earth: 382 000 km (238 800 miles)
Sidereal period (period of revolution about the Earth before reaching the same point in the sky relative to a fixed star): 27 days, 7 hrs, 43 min
Duration of one lunar day: 27 days, 7 hours, 43 min, same as sidereal period
Synodic period (period of revolution about the Earth before reaching the same position in the sky relative to the Sun): 20 days, 12 hours, 44 min
Mass: 1/81 that of Earth (or 0.012 Earth masses)
Density: 1/6 that of Earth
Surface temperature: +100°C daytime; −150°C nighttime

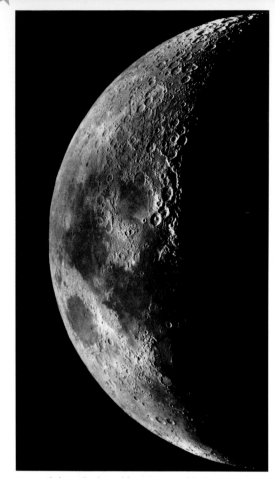

At the start of the first quarter, mount Hemus and the Caucasus mountains are seen bordering the sea of Serenity, as well as the clefts of Triesnecker and Hyginus, and the circular basin Hipparcus. The craters Autolycus and Cassini are visible, as well as the Alpine Valley, looking very much like a giant claw mark on the Moon's surface.

About eight days after the New Moon, some of the most beautiful lunar features emerge. These include the craters Maginus and Purbach, and the Straight Wall a giant fault line about 120 km long. Pay particular attention to the prominent group of craters, Arzachel, Alphonsus, and Ptolemaeus, the Apennine mountain chain, as well as the sea of Rains. The craters Archimedes and Aristillus are also prominent and do not overlook the enigmatic basin Plato.

Clavius is well placed by the 10th day of lunation. One of the most conspicuous lunar features, this gigantic formation is as large as Switzerland! Other prominent features emerging from the shadows at this time are Tycho, Eratosthenes, and Timocharis. Longomontanus, the Carpathian mountains, the Gulf of Iris and the highlands Heraclitus are also well placed. You certainly cannot overlook Copernicus, one of the most beautiful circular craters on the Moon, complete with a prominent central peak.

By the 12th day, the craters Aristarcus and Kepler are well placed, both displaying an extensive system of bright rays. The Moon is full by day 14. Although a spectacular sight to the

Around the 6th day of lunation and before first quarter Moon, be sure to catch the splendid trio of craters, Theophilus, Cyrillus and Catharina, shown here in the center of the photograph. South is at the top as is characteristic of all telescopic images.

THE INFAMOUS MOON ILLUSION

Have you ever noticed that the Moon appears to be much bigger near the horizon than overhead? This enigma has perplexed people since 350 BC. Aristotle thought it was due to atmospheric gases causing a distortion in the Moon's appearance near the horizon. It was not until 1000 years later that an Arabian physicist, Ibn Alhazen, proposed the first plausible explanation. He suggested that when the Moon is low we see it against a background of familiar objects near the horizon (trees, houses, etc). This provides us with a visual "reference point" against which to gauge the size of the lunar disk, a reference lacking when the Moon is overhead.

How can you verify this yourself? Perforate an index card with holes of various sizes and hold it at arm's length over the Moon when it is near the horizon. Match it with the hole of the most appropriate size. Repeat this experiment when the Moon is near the zenith. Surprising, isn't it?

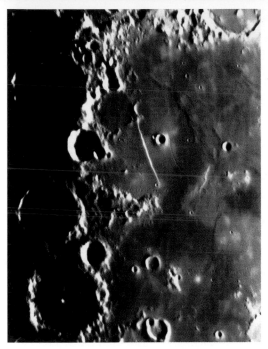

The Straight Wall or Lunar Sword, is one of the most remarkable geologic features on the Moon. It is a fault line over 300 m high and 120 km long.

Archimedes, Autolycus, and Aristillus are visible near the lunar Apennine mountain chain. The valley of the Alps appears near the bottom (south).

naked eye, the Moon is least well positioned for telescopic observation at this stage. Since the Sun's rays strike it face on with no shadows to help highlight surface features. A neutral density filter comes in handy at this point, not only to cut down the glare, but also to help visualize the Moon's extensive system of rays extending from such craters as Tycho, Copernicus, Aristarcus, and Kepler. (These rays represent material ejected and scattered radially from the point of impact when those craters were formed.)

After Full Moon, you can resume observing lunar features but illuminated from the opposite direction, and in reverse phase.

What can you see with various instruments?

◎ With unaided eye
The principal geographic features can be discerned, including the larger oceans and seas, some of the larger ring structures and major ray systems. Major lunar/planetary conjunctions are also beautiful sights, when Venus, Mars, Jupiter, and Saturn may be visible in close proximity to the Moon.

◎ With binoculars
You can easily see many of the principal lunar features first glimpsed by Galileo with his 30-mm refractor, including all major oceans, seas, continents, and craters. Binoculars are ideal for observing Earthshine, particularly during days 3–5 after the New Moon, as well as for lunar eclipses.

◎ With a small 60-mm refractor
The smallest lunar feature you can clearly resolve with an instrument of this size is about 4 km in size. You should have no difficulty spotting the valley of the Alps, the Hyginus Cleft, and Aristarchus, the wave-like

1 Mare Frigoris or Sea of Cold
2 Gulf of Iris
3 Mare Imbrium or Sea of Rains
4 Carpathian Mountains
5 Aristarchus
6 Eratothenes
7 Oceanus Procellarum or Ocean of Storms
8 Kepler
9 Copernicus
10 Reinhold
11 Lansberg
12 Grimaldi
13 Sea of Knowledge
14 Gassendi
15 Mare Humorum or Sea of Moisture
16 Bulliadus
17 Mare Nubium or Sea of Clouds
18 The Straight Wall
19 Pitatus
20 Wurzelbauer
21 Hainzel
22 Schickard
23 Tycho
24 Maginius
25 Clavius
26 Longomontanus

The labeled lunar maps on the following two pages show the detail visible in a good pair of binoculars

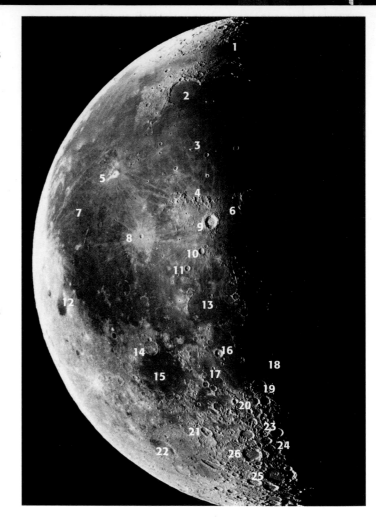

"undulations" in many of the seas, and literally thousands of craters. At 100× the Moon appears as close as it would from only 3 800 km or about 2 300 miles, just as if you were in a spaceship closing in on it.

● **With a 120–150-mm (4-6-inch) telescope** With a telescope of this size, the range of things to investigate on the Moon becomes almost limitless. Countless small craters, rills and other fine detail become apparent inside larger craters and ringed structures. Under really good seeing a 6-inch telescope can resolve objects one arc second and less in apparent size, meaning you will be able to spot craterlets only 1.5 km (1 mile) in diameter.

1 Mare Frigoris or Sea of Cold
2 Aristotle
3 The Lunar Alps
4 Plato
5 Caucasus Mountains
6 Archimedes
7 Mare Serenitatis or Sea of Serenity
8 The Apennines
9 Mare Vaporum or Sea of Vapors
10 Mare Crisium or Sea of Crisis
11 Mare Tranquilitatis or Sea of Tranquility
12 Apollo 11 landing site
13 Albategnius
14 Mare Fecunditates or Sea of Fertility
15 Ptolemeaus
16 Alphonsus
17 Arzachel
18 Theophilus
19 Cyrillus
20 Catharina
21 Mare Nectaris or Sea of Nectar
22 Purbach
23 Regiomntanus
24 Tycho
25 Clavius
26 Mare Australe or Southern Sea

Note that when using a star diagonal with your telescope, images are reversed left to right

An unusual sight, a lunar eclipse

Lunar eclipses are readily observed from the city and pose absolutely no danger to your eyes. They occur during Full Moon and only when the Sun, Earth, and Moon are aligned in a way that the Earth's shadow is projected in space toward the Moon and covers it temporarily. The Moon usually turns a striking copper-red color at such times, due to sunlight being refracted through our own atmosphere, which acts as a prism, or filter that scatters or blocks out other colors. Why is there not an eclipse during each Full Moon? Because the Moon usually passes slightly above or below the Earth's shadow during most lunar cycles.

The Sun: observe it with caution

The Sun is our own star and the only one we can actually observe during the day. We rely on it for light, energy, and our planet's well being; in short we need it for life itself. It is also a fascinating object for study, but we must take some major safety precautions before doing so.

One star among countless others

The Sun is really a pretty ordinary star, neither too large, nor too small; not too hot, and not too cold. In short, it is an average star like billions of others in our own galaxy. Thanks to the Sun, life was able to evolve on Earth but thanks to that as well it will eventually cease in a few billion years, when the bulk of the Sun's fuel has been consumed and it slowly expands into a red giant and engulfs its own system of planets.

Seen up close, the Sun is an enormous sphere of extremely hot gases in a perpetual state of tumultuous activity. Although the Sun is large, with a diameter equal to four times the distance between the Earth and the Moon, there are stars in our galaxy 10,000 times larger than the Sun!

The Sun's primary activity is the conversion

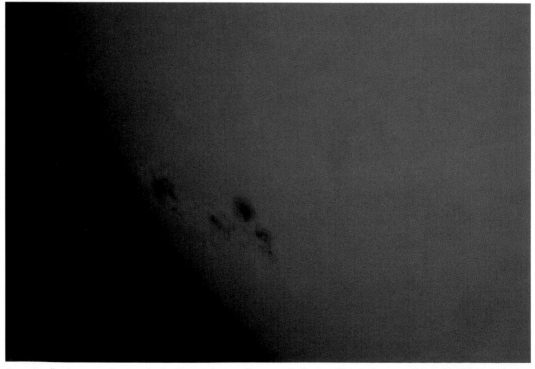

In a small telescope the Sun looks like a giant, yellow gas ball, pitted by dark sunspots and bright, white patches called faculae.

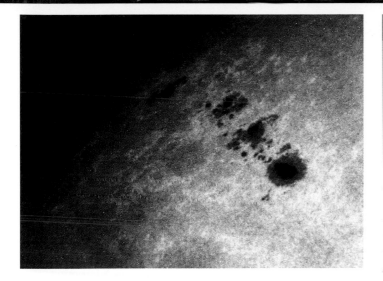

Sunspots are surface features that are slightly cooler than the rest of the photosphere and therefore appear relatively darker. They are centers of powerful magnetic storms.

of hydrogen to helium and eventually to more complex elements (carbon, oxygen, and various metals). In this thermonuclear process, the Sun actually converts over 4 million tons of hydrogen to helium per second! Even at this seemingly torrid pace though, the reduction in the solar mass is only a negligible 5/1000 in 3 billion years. These activities generate an enormous amount of energy. In addition to visible light, such damaging radiation as ultra violet, X-rays and gamma rays is also emitted. At the other end of the electromagnetic spectrum, the Sun emits infra-red radiation (heat) and radio waves. Fortunately our atmosphere filters out much of the more energetic and dangerous radiation.

What can you observe on the Sun?

There is much to see on the Sun with a (fully filtered) small telescope, making it a choice target for urban observers. The visible part of the Sun is called the photosphere. This hot, very luminous layer is about 300 km deep and emits essentially all light in the visible part of the spectrum. The photosphere appears as a yellow (depending on the type of filter used) disk in a small telescope, darkening toward the limb regions, a clear indication that the Sun is indeed a sphere.

The photosphere is the site of much activity, manifested both by sunspots and the bright, often spectacular faculae that seem to pierce the Sun's surface. Sunspots are actually storm-like regions, where intense magnetic fields are generated. Sunspots generally appear in a narrowly defined region on the Sun, between 40° and 50° north and south latitude, and are usually more numerous near the solar equator.

Under high magnification, sunspots can be resolved into two distinct zones, a darker central "umbra" and an outer, more irregular "penumbra." Sunspots often appear in groups of two to five individual spots and can progressively change in shape, size, and appearance from day to day. They generally last a few weeks and then disappear.

Sunspots are "slightly" cooler (between 4000 and 4500°C) than the rest of the photosphere, whose temperature is closer to 6000°C. Despite their dark appearance, sunspots are actually brighter than an electric arc, and only look darker because of their relative contrast against a much brighter background.

In addition to sunspots, you can usually see many brilliant streaks or faculae, which are extremely bright particularly when seen against the relatively darker limb regions of the Sun. Sunspots, are closely linked with faculae and often form at their edges.

Under good seeing conditions and with a high-resolution telescope, the convective nature of the solar surface becomes quite evident. This is manifested through very fine structures in the photosphere called solar granules. Although these structures look deceptively like rice grains or boiling oatmeal, remember that each one of them is about the size of France.

The chromosphere and corona are only visible during a total solar eclipse when the Moon completely covers the Sun's disk. Even during the partial phases (as shown here) these solar features remain invisible.

Solar eclipses, among nature's most grandiose sights

The highest portion of the Sun's atmosphere, called the chromosphere, becomes visible only during a total eclipse, when the Moon passes directly in front of the solar disk and (temporarily) blocks off the photosphere. The chromosphere is more transparent and less luminous than the photosphere and so is normally not visible. During a total eclipse, however, it appears as a rose colored ring, with immense gas extensions called prominences. Beyond the chromosphere we see the corona, a crown-like,

During totality when the Moon covers the Sun's disk almost completely, we can see the outermost portions of the solar atmosphere which are normally invisible.

gaseous envelope reaching temperatures of 100 000°C. This magnificent sight is unfortunately not visible except during an eclipse or with the aid of specialized instrumentation not generally available to amateurs.

FACT AND FIGURES ABOUT THE SUN

Age: 4.5 billion years
Expected additional lifespan: 5 billion years
Diameter: 1 390 000 km (864 000 miles), 109 times diameter of the Earth or 4 times the Earth-Moon distance.
Volume: 1 300 000 that of the Earth
Distance from Earth: 149 597 870 km (93 000 000 miles), 8 min 18 sec for light to travel. This equals 1 AU (Astronomical Unit)
Mass: 332 700 Earth masses (99.9% of the mass of the total solar system)

Density: 1.41 (Water = 1)
Composition: 90% hydrogen, 10% helium (traces of carbon, oxygen, sodium, calcium, iron, etc)
Luminosity: 600 000 brighter than the Full Moon
Period of Rotation: 25.4 days (at equator)
Surface Temp: 5 700°K
Core Temp: 15,000,000°K
Core Pressure: 400 billion atmospheres

The Sun reawakens every 11 years

At certain times the Sun appears devoid of sunspots for periods lasting several weeks. At other times it exhibits a great many spots. The number and size of sunspots, as well as indicators like radio emissions, characterize the cyclical nature of solar activity. In 1834 the German astronomer Schwabe (who observed the Sun for more than 50 years) noticed that the sunspot cycle was about 11 years long (or about 11 years between either maximum or minimum sunspot numbers). The most recent peaks occurred in 1989 and in 2000.

The exact mechanism underlying this cycle of sunspot maximum and minimum is still a scientific mystery. People have tried to correlate these cycles with major wars (1870, 1917, 1938, etc.), frequency of suicides, heart attacks, and even nervous depression! The only established correlations, however, between sunspot maxima and other physical phenomena include interference with radio and satellite communications, generation of magnetic storms, and enhancement of aurora activity, often to the delight of astrophotographers.

How to observe the Sun in complete safety

Solar observing is both easy and enjoyable, but requires exercising some very strict precautions. There are basically two completely safe ways to observe the Sun.

⊙ **The projection method.** This is both the simplest and least expensive way of observing the Sun, requiring no more complex accessory than a cardboard box with a white interior. Using a medium power eyepiece, align your telescope in such away that it projects the Sun's image on to the white cardboard placed behind the ocular, much a like a screen. To help point the telescope properly, move it around until its shadow projects a very narrow silhouette against the cardboard screen. The solar image should appear at the center of that shadow. WHILE DOING THIS, NEVER LOOK AT THE SUN THROUGH THE TELESCOPE'S FINDER. In fact, leave the finder's disk cap on to prevent this and to protect its internal reticle.

A number of commercial suppliers sell small solar projection screens that can be solidly attached to the telescope tube or eyepiece assembly or to project an image through a star diagonal. The main advantage of this type of set up is that several people can

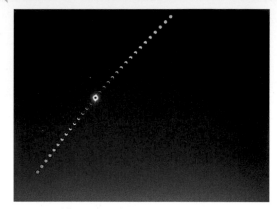

It is possible to record every stage of a total solar eclipse on one frame by taking one exposure every six minutes with a camera that has multiple exposure options.

observe at the same time. Solar projection methods are only recommended for 60 mm refractors and simple eyepieces, but even then a full aperture filter is both safer and less expensive than a quality eyepiece.

○ **Full aperture filters.** Many small department store telescopes are supplied with so-called "Sun Filter" accessories that are designed to be placed in front of an eyepiece. Filters like these are ABSOLUTELY NOT RECOMMENDED. First, since they are placed near the telescope's focal plane, they can easily overheat and shatter, exposing your eye to the full intensity of the Sun, and, second, many of the filters are cheap, darkly stained glass which can still transmit infra red light and cause gradual damage to your eyes.

Much more effective and totally safe are full aperture filters placed at the front end of your telescope. Several quality brands are available. They are usually made of optical quality glass, covered with a fine film of nickel-chromium alloy of optical density 5, and only transmit about 1/100 000 of the Sun's light. This type of filter is the only totally safe filter both for your eyes and your optics.

Full aperture glass filters of optical density 4 are known as "photographic" filters and

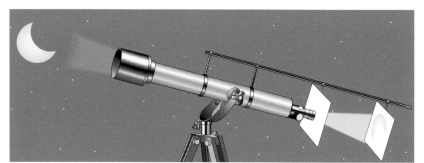

The solar screen method for looking at the Sun has the advantage of being easy and inexpensive and allowing several people to observe simultaneously.

An extremely simple way to observe the Sun is to poke a pin size hole in the bottom of a shoe box and project the image on to the inside of the cover. This is the same principle of the classic "pin hole" camera.

transmit about 1/10 000 of the Sun's light. These should not be used visually but only for solar photography with an additional eyepiece pre-filter. Jim's Mobile Inc. (JMI), Orion and Thousand Oaks are three American suppliers of full aperture glass filters like these and they are available for virtually every type of telescope on the market. They are priced between $60 and $150, depending on size and quality (e.g. ~$60 for a Meade ETX 90).

There are also several Mylar type and other coated plastic solar filters available for this purpose. These are usually extremely thin, aluminum-coated materials, which can be wrapped around or attached to the front end of the telescope. They are usually not of the same optical quality as glass filters and are really designed for the occasional solar observer only. It is important to note that you must not use ordinary aluminized wrapping or similar plastic material as solar filters, only filters commercially sold for that purpose. Aluminized wrapping materials are not of good optical quality and may transmit UV and other harmful rays which can damage your eyes.

To point at the Sun with full aperture filters align you telescope and follow the method described above for projection screens.

A full aperture filter provides the best optical definition and the safest way of observing the Sun. A filter from Thousand Oaks Optical is shown here with a Meade ETX 90 telescope.

What can I see with various instruments?

○ With the naked eye
Except for eclipses and with the aid of special filter eyeglasses, there is really little of interest to see on the Sun with just your eyes.

○ With binoculars
You can observe the Sun with binoculars equipped with full aperture filters or properly aluminized plastic material, although never with just ordinary sunglasses. About all you will see, however, are the larger sunspots looking like tiny dots. You can use binoculars with solar projection and see a little more detail.

○ With a telescope
With a full aperture filter or projection methods (but only with small refractors) you can readily observe details in individual sunspots and sunspot groups, their daily numbers and position, as well as changes in their appearance over time. The larger faculae will also be visible and with larger instruments, solar granulation can also be resolved.

The planets: siblings of the Sun

The Sun was born out of an immense, spinning disk of gas, and dust, which then gave rise to our system of planets. All the elements and primordial materials we are familiar with today were integral components of that initial cloud of gas and dust, and the planets are indeed the Sun's siblings.

Our solar system

All the planets revolve around the Sun in the same direction and overall pattern. Seen from the Earth however, this perspective is quite different since the planets move around the sky near the **ecliptic** (the apparent path of the Sun, see page 14) in a relatively narrow circle known as the Zodiac. Consequently you will always have to look among those constellations to find them. The planets revolve around the Sun in elliptical, rather than circular, orbits, in a counterclockwise direction.

Although star-like in appearance to the naked eye, the planets are readily

LEVELS OF INTEREST
T = telescope; **B** = binoculars
1 **T** (or **B**) = of no interest
2 **T** (or **B**) = somewhat interesting
3 **T** (or **B**) = peaks the curiosity
4 **T** (or **B**) = very interesting
5 **T** (or **B**) = not to be missed

Illustration of the solar system showing the relative sizes of the planets (and the Sun). Only a proportional indication of their distances from each other is indicated since these could not be drawn to scale here. Even when all the planetary masses are combined, they amount to less than 1% the mass of the Sun.

distinguished from stars, because most are quite bright and, being much closer to us, do not twinkle. If you watch the position of any planet carefully for a few days or weeks, you will soon notice that they slowly wander amid the fixed stars. This characteristic was noted by the earliest astronomers and earned them the name *planet* or wanderer in Greek.

During almost any lunation, the Moon will appear to pass near or even close to any planet that happens to be visible at the time. This is of course only an apparent close encounter, since the Moon is much closer to us than any planet. Events like these are called conjunctions and are usually listed in the ephemeredes and magazines.

A guided tour of the planets

Moving outward, the planets orbit the Sun in the following order: Mercury, Venus, Earth, Mars, Jupiter, Saturn, Uranus, Neptune, and Pluto. A collection (or belt) of minor planets or asteroids is also found between the orbits of Mars and Jupiter, with Ceres being the largest at approximately 1000 km in diameter. Anyone interested in observing and following asteroids with a telescope is hampered by the fact that they are only distinguishable from background stars by their slow movement.

The nine major planets fall into two distinct groups, due mainly to their very different physical characteristics. Planets that most resemble the Earth in terms of their relatively small size and dense, rocky composition are called terrestrial planets and include Mercury, Venus, Earth, and Mars. The others known as **jovian** or **giant** planets, resemble Jupiter and are much larger and less dense than the Earth-like planets and are also composed primarily of gases. They do not contain a solid core as such, and include Jupiter, Saturn, Uranus, and Neptune. Pluto is not well understood but is probably more terrestrial in nature.

Mercury TT BB

Mercury and Venus are called the **inferior** planets because their orbits are inside ours (and closer to the Sun). For that reason these planets are never visible in the middle of the night, but only around sunset or sunrise depending on whether their elongation is east or west of the Sun.

Mercury is very difficult to observe due to its proximity to the Sun. You can try to catch it in binoculars during periods of maximum elongation; at best it will rise about 2 hours and 15 min before the Sun or set 2 hours and 15 min after it. At such times, it will be visible for a several days and resemble a fairly bright star suspended in the evening haze near the horizon, or close to the horizon at twilight. When Mercury is favorably placed you can observe it for a couple of weeks before it is again lost in the glare of the Sun. Its period around the Sun is actually quite fast, since one year on Mercury is only 88 days long.

Due to its small size, 4880 km (3200 miles), just barely larger than the Moon, Mercury reveals no real surface detail and certainly its many craters cannot be resolved telescopically. Under high magnification, the best you can hope for is to see its small, featureless crescent.

Venus TTTT BB

Venus in many ways is the Earth's sister planet, both in terms of size and surface relief. That is where the similarities end however, since beautiful Venus is wrapped in a very dense atmosphere and is covered with sulfuric acid clouds, which completely hide its surface from view.

Like Mercury, Venus appears near the western horizon at dusk (and so is often the first "star" visible after sunset), or near the eastern horizon at dawn. Due to its extreme brightness, it is impossible to confuse this planet with a star, none of which come close to it in magnitude. Venus is easy to find and follow in the sky for several months each year.

If your telescope has an equatorial mount (and is polar aligned) point your telescope at Venus before sunrise and (since it is so bright) you will be able to follow and observe it in full daylight.

Although striking to the eye, Venus unfortunately is often disappointing through binoculars or a telescope, since its bright mantle of clouds prevents you from seeing any of its surface features. Only its moon-like phases are readily apparent, and change progressively as it changes position relative to the Sun. Just before inferior conjunction, it will show an extremely thin crescent with extended cusps. With a telescope and just 40×, Venus will look as big at that time as the Moon does to the naked eye.

Venus is often visible as an extremely bright (and featureless) crescent. This photograph was taken with a 60-mm refractor at 840-mm focal length.

Mars TTTT B

For nearly two hundred years this desolate little world has stirred human passions and imagination like no other. More than one astronomer has dedicated their entire career to the study of Mars. The wealthy American amateur, Percival Lowell, actually built an entire observatory in the late 19th century, for the sole purpose of studying the famous "canals" of Mars discovered earlier by the Italian astronomer Schiaparelli. Although these canals proved illusory in the end, they nevertheless gave rise to the belief that Mars was inhabited by intelligent beings.

Do not expect to see the canyons, volcanoes, and dry riverbeds on Mars either. Those have been imaged by the many spacecraft sent to this planet, while also showing that this small neighboring world is mostly arid desert and only has a very tenuous atmosphere. Water, which once played a major role in shaping the early geology of Mars, is no longer flowing freely on its surface because of its very low atmospheric pressure (7/1000 that of Earth). The Martian surface is much more accentuated than ours, however, and includes giant volcanoes like Olympus Mons, which has a base diameter of 600 km (nearly 400 miles) and is more than 24 km (16 miles) high!

Mars appears as a small red-orange disk in the telescope, a color due to the presence of large quantities of oxidized iron in its soil. With magnifications of 150–200×, you will be able to see the principal albedo markings on Mars, including dark and light features. These are really visible only during favorable oppositions, however, when Mars and Earth pass in relatively close proximity to each other. Look particularly for a dark, triangular shaped feature called Syrtis Major, and a prominent feature, Mare Acidalium, not far from the Martian North Pole. You can also follow the seasonal changes in the size of the Martian

Color filters can come in handy when observing the planets. Use a red or orange filter to accentuate the dark features on Mars, and a blue filter to highlight the polar caps and atmospheric phenomena like clouds. Try using color filters on the other planets as well, particularly with Jupiter to enhance contrast and to better differentiate its many cloud bands and the famous Red Spot.

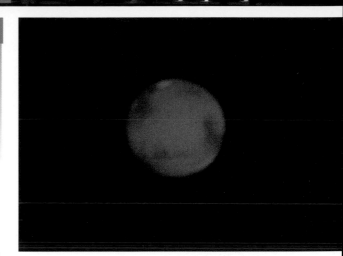

polar caps, which are composed mostly of frozen carbon dioxide and look like minuscule white patches in the telescope.

Mars has two small satellites, Phobos and Deimos. They are of irregular shape and much too small to be seen with most amateur size telescopes.

Jupiter TTTT BBBB

Often nicknamed the "amateur's planet," Jupiter is truly a favorite among observers. Due to its enormous size, it appears as large in a telescope at 40× as the full Moon does to the naked eye. You certainly will not have any trouble finding it since it is second only to Venus in brightness, with the added advantage that it is often visible for long periods of time.

Located well beyond the orbit of Mars, about 778 million km (486 million miles) from the Sun, Jupiter is the largest planet in the solar system. It is an enormous sphere of gas some 142 800 km (nearly 89 000 miles) in diameter and composed primarily of hydrogen and helium, as well as ammonia, methane, and water vapor. Due to intense internal activity, this gas giant actually emits more energy than it receives from the Sun.

Jupiter's core is thought to consist of extraordinarily compacted hydrogen, in a liquid/metallic state harder than steel! If Jupiter's mass were large enough to initiate internal thermonuclear reactions, it would actually ignite into a star. We would then be part of a binary system with two suns in our

Don't look for the infamous "canals" on Mars; all you can see is some of the surface albedo (brightness) features (under good viewing), but these too can be poorly defined in small telescopes. However, when imaged with a powerful telescope, as shown in this photograph taken with the 1-m (40-inch) instrument at pic du Midi Observatory, Mars reveals a lot of subtle surface detail.

Giant Jupiter is a favorite target of astrophotographers since it is relatively easy to image.

THE "GALILEAN" SATELLITES

Although Jupiter has (at least) 36 moons only four are visible in small telescopes. These satellites were discovered in 1610 by Galileo and are named: Io, Europa, Ganymede, and Callisto. Hardly a night goes by when you cannot observe one or more of these moons either in transit across the face of Jupiter or passing behind it. You can often see the dot-like shadow of a moon in transit projected on to the disk of the planet.

sky. You can consider Jupiter a "failed" star, which is clearly a good thing since we might not be here otherwise!

With a magnification of 100–150×, Jupiter's disk appears slightly "flattened" at the poles and traversed by several parallel cloud bands, alternately darker and lighter in color and intensity. If you are patient enough you will also catch the famous Red Spot, whose contrast and appearance changes from year to year. Jupiter's atmosphere is very active and in a perpetual state of turmoil, and this is reflected in the continuing changes in the cloud band details and colors.

Saturn TTTTT BBB

There is almost total agreement that Saturn is hands down the most beautiful object in the solar system. Nothing compares to seeing it "live" in a telescope for the first time. It seems that no photograph or illustration can do it justice. To many people that first view of Saturn is more than an unusual sight, it is an emotional experience! No astronomer ever forgets that moment.

Even though Saturn is smaller than Jupiter, it bears much similarity to it and is also a giant gas planet located about 1.5 billion km (888 million miles) from the Sun. What distinguishes it from all other planets, however, is a spectacular system of rings clearly visible with even the smallest telescope. Saturn's ring system is actually a collection of several rings, circling the planet's equator,

composed of left over debris of rocks, ice and dust, and separated by a number of gaps or "divisions", which are actually areas less densely populated with debris. A small telescope will readily show Cassini's division as a dark zone separating the two principal ring components. Encke's division and the innermost "crepe" are harder to see and require at magnification of at least 200×.

Saturn's rings appear to undergo a progressive change in visibility, from wide open to almost invisible. This is caused by a 27° tilt in the angle of inclination of Saturn's equatorial regions to the plane of its orbit, changing our line of sight as the Earth passes above or below this plane. This happens twice in the course of one Saturnian year (29.5 Earth years), as we see the rings from above, from below, and edge on.

THE SATELLITES OF SATURN

Saturn has at least 18 satellites. Among these, Titan is by far the largest and most easily seen in a small telescope. Rhea, Iapetus, Tethys, and Dione are considerably fainter and require at least a 100-mm (4-inch) telescope to be visible.

Saturn is undoubtedly one of the most beautiful sights in the sky and one that no one is likely to forget.

Left: The planet Uranus, the only one discovered by mathematical computation, and well before the invention of calculators and computers. Center: Neptune, another gas giant, composed of (mostly hydrogen and helium) and traces of methane and ammonia, imaged here by the Voyager probe and showing a great amount of fine detail. Right: Pluto, a planet essentially impossible to observe from the city.

Uranus, Neptune, and Pluto T B

We now find ourselves at the outermost limits of the solar system. At this distance, the Sun appears as a very bright star among the rest, with a correspondingly weaker output of heat energy. Uranus, Neptune, and Pluto are of no great interest to amateur astronomers, being difficult to observe even with very powerful telescopes.

○ Uranus was discovered (by accident) in 1781 by William Herschel, an English amateur astronomer who built some of the largest telescopes of his time. At first he thought he had discovered a comet, but its tiny disk confirmed it as the 7th planet of our solar system. Like Saturn, Uranus has a ring system and rotates about its axis in a retrograde direction, with its axis curiously tilted in the plane of its orbit. Hard to find against the background of stars, Uranus' tiny blue-green disk is best seen with at least a 200-mm (8-inch) aperture telescope.

○ Neptune was discovered in 1846 by the German astronomer J.G. Galle, based on predicted positional calculations by the French astronomer d'Urbain Le Verrier and by English astronomer John Adams. Based on observed perturbations in the motion of Uranus across the sky, both astronomers predicted to within 1° the position of the unknown planet. Neptune is even more difficult to observe than Uranus and presents only a minuscule disk, never larger than 2 arc seconds nor brighter than 8th magnitude. It is just a blue dot among the stars and not much more in a small telescope.

○ The last known planet in our solar system, Pluto, was discovered photographically in 1930 by American astronomer Clyde W. Tombaugh. Smaller in diameter than the Moon, Pluto never comes closer to the Sun than 4 billion km (2.7 billion miles) and orbits as far as 7 billion km (4.6 billion miles). At 14th magnitude, you can just glimpse it under very clear, dark skies in a 400 mm (14-16-inch) telescope. In other words, forget it from the city! Pluto has one known satellite, Charon. Discovered in 1978, its diameter is so close to that of Pluto that they are considered binary planets, Pluto–Charon, unique in our solar system.

Comets, evening visitors

Jonathan Swift wrote, "Old men and comets have been venerated for the same two reasons: their long beards and their pretense of foretelling the future." Yes, these celestial vagabonds have always been associated with fearful superstitions.

Unexpected visitors

Comets have historically been viewed with fear and apprehension. Why might that be? Maybe it is because they did not seem to abide by the rules that other astronomical objects follow. The Moon, the Sun, the planets, and the stars all appear and move in ways that are predictable, regular, stable, and somehow reassuring. But not so with comets. They appear out of nowhere, and in so doing cause "disorder" in our seemingly ordered universe.

It is not surprising therefore that the astrologers of old saw them as messengers or omens, usually bringing bad news or announcing imminent catastrophes: the death of a famous person, or, worse, another war. No wonder people were frightened of comets.

Stranger still, some comets made an appearance and were never seen again, while others returned periodically. Sir Edmund Halley, the English astronomer royal first proved that in 1705. He demonstrated that the comets observed in 1456, 1531, 1607, and 1682, were one and the same, and predicted its return in 1758, alas, a few years after his own demise. Halley's comet last appeared in 1910 and 1986, and is awaited again in 2062.

What exactly are comets?

Imagine an enormous, dirty snowball, several kilometers in diameter, entirely composed of ice, dust, and rock fragments. That pretty much describes a comet. As it approaches the Sun, this nucleus of ice brightens enormously as it begins to out gas some of its volatile contents into space, forming a diffuse coma around itself and a growing tail in its wake. Comet tails, consisting entirely of gas and dust, can attain several million kilometers in length. These trails of gas and dust always point away from the comet nucleus in a direction opposite that of the Sun. As they get closer to the Sun some comets take on an unexpected and striking change in appearance, a fact that no doubt contributed greatly to our ancestors' misgivings about them.

Cometary nuclei and the associated coma can be observed to great advantage with a telescope. However, binoculars are definitely the instruments of choice here, since only they have the wide field of view necessary to see the full extent of the comet's tail across the sky. This can be a truly spectacular sight. Like the planets, comets are also part of the solar system. Their orbits, however, are generally highly elliptical, either parabolic or hyperbolic, in which case the comet will pay us only a short visit never to be seen again. Even periodic comets can take several thousand years to return, particularly if their orbits are very elongated.

Where do they come from?

The exact origin of comets remains a mystery. The most current view is that they were formed at the same time as other components of the solar system about 5 billion years ago. They are thought to be part of the Oort cloud, a ring of primordial comet-like material about one light year out from the Sun. Every once in a while, orbital instabilities and collisions

Despite some very heavy light pollution from nearby Paris, this photograph nicely captured comet Hale-Bopp over the city of Saint-Cloud in France.

You do not have to be a great astrophotographer to capture a brilliant comet on film. This is a one-minute time exposure with camera firmly mounted on a tripod.

among Oort cloud comets, will knock one out and in the direction of the Sun.

Over the ages, many highly dedicated amateurs have hunted for comets from dark locations and with appropriate optical equipment. Often they have done this to the exclusion of other astronomical observing, and always with the hope of finding a new comet for posterity. New comets are always named after their discoverers. Their ranks include some well-known names: West, Encke, Bennet, Austin, and Kohoutec, and, more recently, Shoemaker/Levy, Hyakutake, and Hale-Bopp, all discovered by very happy amateur astronomers.

Meteors

Often called "shooting stars," meteors actually have little do to with the stars proper. We see meteors because periodically the Earth crosses the path of an ancient comet, either a periodic comet or one now long gone. The dusty remnants of such comets still orbit the Sun, however, and enter our atmosphere as we pass through this wake. Most such particles range in size from a few microns to millimeters. Meteors appear as flashes because they travel extremely fast (up to 100 000 km/hr) and burn up in the atmosphere very quickly. Some meteors leave spectacular trails (and occasionally an explosive sound). Meteors also come in so-called "showers." The effect is somewhat like traveling through snowflakes in a car at night, where everything is coming at you from one direction. The same happens during a meteor shower, when most meteors appear to come from a defined point in the sky called the **radian**. Meteor showers are named after the constellation from which they seem to come (e.g. Leonids, Perseids, etc.) and can occasionally be seen even from darker urban locations.

The stars above our roof tops: season by season

The spring sky

In its trek around the Sun, the Earth takes us on an celestial tour that changes day by day. Each month new constellations appear in the eastern sky while others set in the west, a reflection of the seasonal cycle we go through annually.

Diving into the deep-sky

Trying to seriously observe the stars under city skies can be quite a challenge. It is certainly not an easy thing to do and often requires much patience. Things can be worse for people

living in really humid areas, close to the ocean, for example, where spotting even the brightest stars can seem impossible at times. Obviously, this will be less of a problem if you live in mountainous areas or on the prairies, where the air is generally drier and less stagnant. Either way, the spring is usually a good time for urban stargazers, since the nights can be cool and transparent in the city, and not as heavily polluted by smog or haze. Before you decide to go out observing, however, check out the clarity of the sky near the horizon and consider setting up after midnight when light pollution tends to be lowest.

Just to reassure you, from someone who has been practicing this hobby for some 30 years now, in and around Paris, the majority of objects I am about to describe (star clusters, nebulae, double stars, etc.), have all been observed (and some even photographed) from deep within the city with nothing larger than 150-mm (6-inch) telescope. For the most part too, we will focus attention on the brighter Messier objects, since with a few notable exceptions, most NGC objects require larger instruments and darker observing sites.

The key to the northern sky

Ursa Major or the Bid Dipper is key to finding your way around the spring sky. It is easy to

M 51 is one of the most observed and popular galaxies in the sky, pictured here in an amateur photograph.

recognize and almost straight overhead, in northern and mid-northern latitudes at this time of year, and helps you locate most of the other interesting constellations using the method outlined earlier. Ursa Major is actually a very large constellation and extends way beyond the seven stars of the Dipper.

Direct your binoculars toward Mizar, the second star (ζ) in the "handle" of the Dipper, and you will see a splendid pair of stars, with Alcor as the other member. Mizar itself is also a double star when observed with a telescope. The name Alcor means "tester" in Arabic, since it was reputedly used by a sultan to test the eyesight of his soldiers.

Moving away from the handle of the Dipper (toward the east), you will find the bright star Arcturus in Boötes, which is close to the beautiful Northern Crown, Corona Borealis. Below Arcturus you will find Spica, the brightest star in Virgo. Moving directly south from the Big Dipper, you cannot avoid Leo, the lion, whose head stars look like a question mark in reverse. Cancer lies a little west of Leo, but is very difficult to see from the city since it is rather faint. Cancer looks like an inverted "Y" which points to one of the treasures of this region of the sky, the large open cluster (M 44) the Praesepe or "Beehive" cluster. This is a splendid object in binoculars.

Slightly south (and west of Cancer) you will find Canis Minor, the small dog, whose brightest star Procyon lies close to the feet of Gemini, the twins, whose names (and

SOME HINTS ON USING THE CIRCULAR STAR CHARTS (FOLLOWING PAGES)

These circular star charts were drafted for the average latitude of France (45°), but are suitable for mid-northern latitudes anywhere around the world. The circular border of these charts represents the horizon and the center shows what lies straight overhead. The cardinal points shown are intended to help orient the chart relative to the direction you are facing when observing.

Hold the chart directly in front of you, turning it so that it points in the direction you are facing, toward the east for example, making sure you do not have east and west reversed. With a little practice, you will soon discover your particular observing "window," in relation to any nearby buildings and other landmarks that may be blocking some sections of the sky.

You will normally see fewer stars with the naked eye from city locations than are shown on these charts. That is to be expected (because of light pollution and other factors). Under dark skies in the countryside, however, you should readily see all the stars and constellations indicated on these maps. The charts without labels on the right-hand page faithfully reflect the colors and relative brightness of the stars and constellations depicted. These will help you gauge the clarity and transparency of the atmosphere at any given time, while the labeled star charts will aid with their identification.

principal stars) are Castor and Pollux. Virgo, Leo, Cancer, and Gemini are four constellations of the zodiac and lie along the ecliptic. You might occasionally see a bright "star" along that path which you cannot identify on your star charts. Could that be a planet? Indeed, it could!

FIVE IMPORTANT RULES FOR URBAN ASTRONOMERS

1 Block off any direct light sources with a tarp, cardboard shield, umbrella, etc., stretched or held in place temporarily while observing.
2 Let your eyes become dark adapted for at least 15 minutes.
3 Check out sky conditions before starting, and avoid clouds, smog, and haze as much as possible for deep-sky observing, although these factors are of less concern for lunar and planetary work.
4 Avoid observing on evenings around Full Moon.
5 If at all possible, observe during the second part of the night (after midnight), when most people go to sleep, and many city lights are turned off (skyscrapers, monuments, car lots, etc.), which can gain you 1-1.5 magnitudes more.

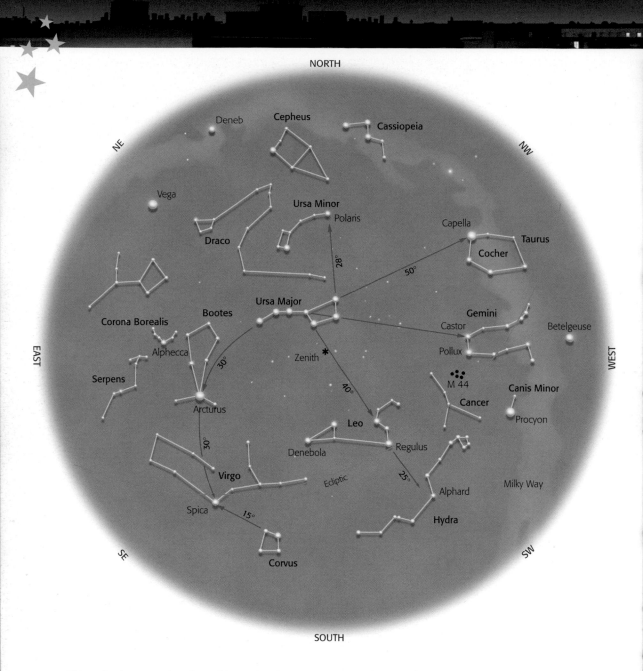

NORTH

NE

NW

Deneb
Cepheus
Cassiopeia

Vega

Ursa Minor
Polaris

Capella

Draco

28°

50°

Cocher

Taurus

Corona Borealis

Bootes

Ursa Major

Gemini

Betelgeuse

Castor

Alphecca

30°

Zenith ✳

Pollux

Serpens

EAST

WEST

Arcturus

40°

M 44

Cancer

Canis Minor

Procyon

30°

Leo

Virgo

Denebola

Regulus

Ecliptic

25°

Alphard

Milky Way

Spica

15°

Hydra

SE

Corvus

SW

SOUTH

This springtime star chart is usable on the dates and times indicated opposite. Note: all times indicated are in universal time (UT)
These charts apply on the following dates and times:

- March 1 at 0 h
- March 15 at 23 h
- April 1 at 21h
- April 30 at 20 h
- May 15 at 19 h

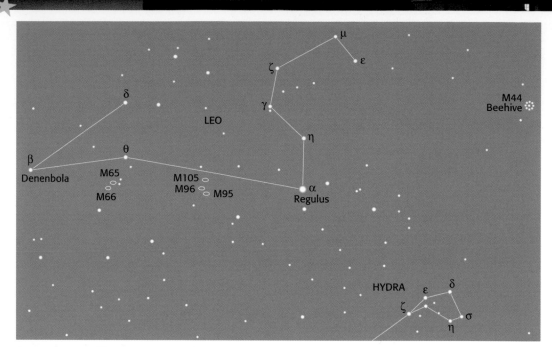

Leo and surrounding area.

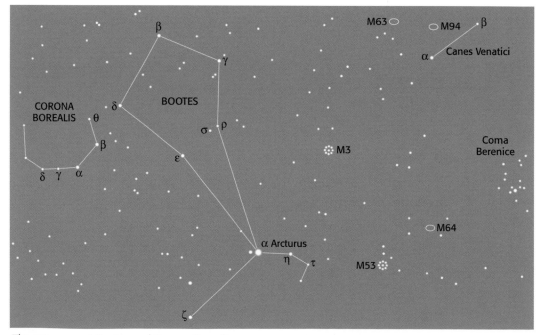

The area around Bootes and Corona Borealis.

Interesting objects in the spring sky

Object	Name	Constellation	Location	Magnitude	Type of Object	Interest/Difficulty
M 44	Beehive	Cancer	near γ Cancri	3.7	open cluster	3B; 1T
M 67		Cancer	2° right of α Cancri	6.1	open cluster	3B; 1T
M 64		Coma Berenices	10° north of ε Virginis	8.8	spiral galaxy	1T
M 3		Canis Venatici	between Arcturus & Cor Caroli	6.4	globular cluster	3B; 1T
α Canun Venaticorum	Cor Caroli	Canis Venatici	south of η Ursa Majoris	2.9–5.4	double star yellow/violet	4 T; 1B
M 94		Canis Venatici	3° north of α Canum Venaticorum	8	spiral galaxy	1T
M 51	Whirlpool	Canis Venatici	3° south of η Ursa Majoris	8.1	spiral galaxy	2T
ζ Ursa Majoris	Mizar	Ursa Majoris	2nd star in Big Dipper	2.4–4.0	double star white/blue	4T; 1B
M 82		Ursa Majoris	near M81	8.7	Irregular galaxy	3 T
M 101		Ursa Majoris	3° north of η Ursa Majoris	8	spiral galaxy	2 T
M 81		Ursa Majoris	between γ & α Ursa Majoris	7.9	spiral galaxy	3 T
γ Leonis	Algieba	Leo	below Regulus	2.2–3.4	orange doublet	5 T
M 65		Leo	7° southwest of β Leo	9.3	spiral galaxy	1 T
M 66		Leo	near M65	9.1	spiral galaxy	1 T
γ Virginis	Porrima	Virgo	3° east of Spica	3.6–3.6	yellow doublet	2 T

The numerical scale from 1–5 = very difficult and low interest to very easy and interesting to observe. Note that galaxies require exceptionally clear and transparent skies to be visible from the city.

B = Binoculars

T = Telescope

The summer sky

The summer sky offers an entirely new panorama of stars. Unfortunately, the nights are short and twilight lasts longer at this time of year, meaning you have to stay up later or get up very early to enjoy the view.

The summer triangle

Look in a southerly direction and raise your eyes almost overhead and you will see an immense triangle in the summer sky formed by three very bright stars, Deneb in Cygnus, Vega in Lyra, and Altair in Aquila. These three constellations located in the heart of the Milky Way form the Great Summer Triangle, the key to the sky this season. In the middle of this triangle and equidistant from its three corner stars lies Alberio. This magnificent orange and blue doublet, makes for a fine view in binoculars and small telescopes. Cygnus is also home to several nebulae, including NGC 7000, the North America, and Pelican nebulae just left of Deneb, as well as NGC 6992 and NGC 6960, the well-known Veil nebula, all unfortunately almost invisible from city skies. Nonetheless, traversing the

M 57 the Ring Nebula in Lyra, remnants of a stellar explosion.

length of the Milky Way with binoculars from Cygnus to Sagittarius is truly a rewarding experience.

Moving slightly west and left of Vega, you will find a star that appears double in binoculars, ε 1 and ε 2 Lyrae. Each of these turns out to be a doublet star as well seen in

---

STAR HOPPING

The simplest way of finding objects not readily visible to the unaided eye is by "star hopping." This technique is used by many amateurs lacking equatorial mounts, as with Dobsonians, for example (although many do it even with other types of telescope mounts, both to learn the sky better and because it is an easy method of locating objects). The method basically involves locating some bright stars in the sector of sky you are interested in, and then "jumping" from star to star with the help of a chart until you find the object you want to observe. It helps having a wide field of view when doing this and clearly binoculars are very useful in that regard. If your Dobsonian is not supplied with a finder, we suggest you get one for it (the Telrad type of locator is favored by many Dob owners for that reason). It is also important to know or have inscribed on a reticle the actual field of view of your binoculars or finder scope so you can use star charts more effectively when star hopping.

Slightly west of Lyra you will find Hercules, the giant, a distinctive "H-shaped" constellation, which is rather difficult to make out from the city. You can pick out the two globular clusters in Hercules, M 13 and M 92 with binoculars, but better still would be a GO-TO equipped telescope. These two magnificent clusters contain (an estimated) million stars each. With a 100-mm (4-inch) or larger telescope you can begin to resolve individual stars in both objects.

---

a small telescope. This is a rare example of a double double! The constellation Lyra forms a small rectangle, whose lower two stars, γ and β frame M 57, the famous Ring Nebula. This smoke-ring like structure is a planetary nebula resulting from a nova explosion. With binoculars it looks just like a fuzzy star, but more interesting detail can be seen with a 100–120-mm (4–5-inch) telescope at 75×.

Toward the center of the galaxy

If you turn back to Altair (α Aquilae) and then move up toward Alberio, you will find a small, Y-shaped constellation called Sagitta (the arrow). You will find M 71 a small open cluster often mistaken for a globular, between γ and δ Sagittae. Just beneath that is M 27, the Dumb-Bell Nebula, one of the finest planetaries in the tiny constellation Vupecula (the little fox). This remarkable planetary is all that remains of an exploded red giant.

Continuing downward through Aquila, we encounter M 11 a splendid open cluster in Scutum, popularly know as the Wild Duck cluster due to its resemblance to a duck in flight. Moving further south still, we find two well-known constellations of the zodiac, Sagittarius and Scorpius. We are now looking right at the center of our own galaxy. The Milky Way is especially dense in stars here and rich in nebulosity, and so full of clusters and other diffuse objects that it is easy to get lost in it all. Unfortunately, these two constellations are situated very low on the horizon at northern latitudes. Still, it is worth your while looking for such splendid objects as M 8, the Lagoon Nebula, and M 20, the Triffid Nebula, both in Sagittarius, and the cluster M 4 near Antares in Scorpius, as well as the splendid cluster NGC 6231 near ζ in the same constellation.

The dense star regions of Sagittarius as you will never see them from the city. A 7-minute time exposure was needed to capture this portion of the Milky Way.

AVERTED VISION

In order to better see dim or diffuse objects like nebulae and galaxies, many observers use a technique called "averted vision." When looking at any object (under normal illumination), its image is focused on to an area of the retina called the fovea, a region rich in cells sensitive to bright light. The fovea is almost blind, however, at very low levels of illumination, and very dim objects are actually perceived better in areas adjacent to the fovea. Consequently, it is actually better when observing a very dim object to not look at it directly, but to avert your vision slightly to one side or below. Since the cones in your retina are also not sensitive to dim light, you will not see any colors either under those circumstances. So instead of seeing the beautiful colors of nebulae recorded on photographs, they will always appear rather monochromatic to the eye, even when observed with large telescopes.

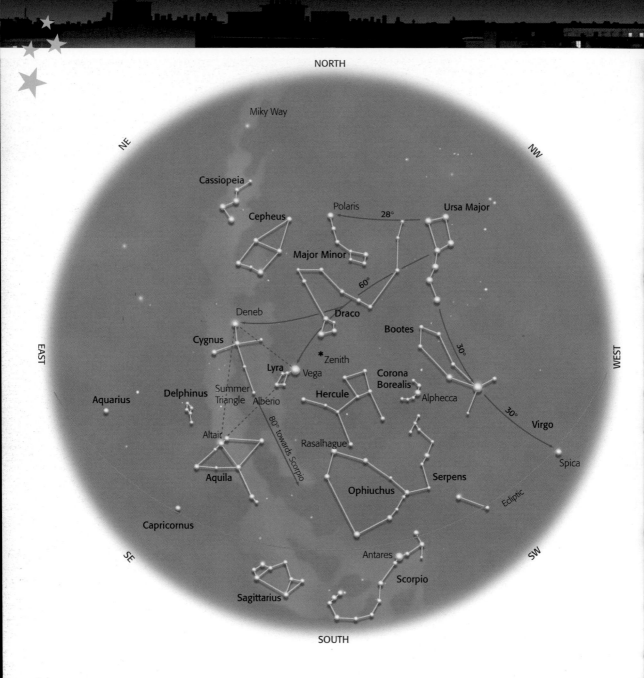

NORTH

Miky Way

NE

Cassiopeia

Cepheus

Polaris   28°   Ursa Major

Major Minor

60°

Deneb

Draco

Cygnus

Bootes

Zenith

Lyra   Vega

Corona
Borealis

30°

Hercule

Alphecca

Delphinus   Summer
Triangle   Alberio

Aquarius

Rasalhague

Altair

30°

Virgo

80° towards Scorpio

Aquila

Serpens

Spica

Ophiuchus

Ecliptic

Capricornus

Antares

EAST

WEST

SE

Scorpio

SW

Sagittarius

SOUTH

This summer sky chart can be used at the dates and times indicated opposite. Note: all times given are in universal time (UT) and you must compensate for differences due to daylight saving time.

These charts apply on the following dates and times:

- July 1 at 23h
- July 15 at 22h
- August 15 at 20h
- August 31 at 19h

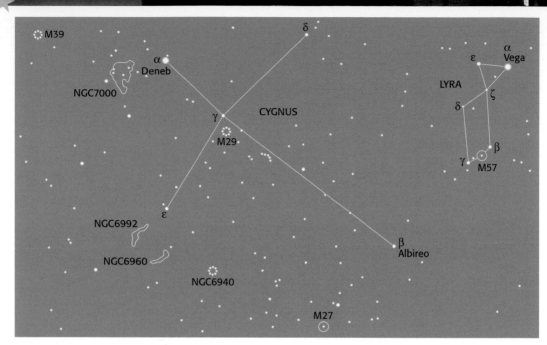

The area around Cygnus and Lyra.

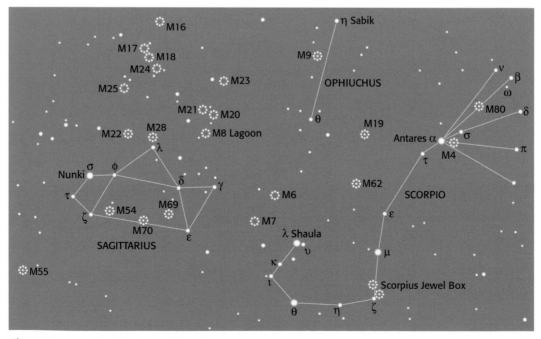

The areas around Sagittarius and Scorpius.

Interesting objects in the summer sky

Object	Name	Constellat.	Location	Mag.	Type of Object	Interest/ Difficulty
β Cygni	Alberio	Cygnus	south of Deneb	3.1–5.1	double star	4 T
61 Cygni		Cygnus	5° southeast of Deneb	5.5–6.4	orange double star	2 T
γ Delphinis		Delphinus	left of Altair	4.5–5.5	yellow-green doublet	3 T
M 11	Wild Duck cluster	Scutum	4° east of Aquila	6.3	open cluster	3 T
M 71		Sagitta	between γ & δ Sagittae	5.5	globular cluster	2 B; 2 T
M 13	Hercules cluster	Hercules	between η & ζ Herculis	5.7	globular cluster	4 T
α Herculis	Ras Algethi	Hercules	southwest of Vega	3.5–5.4	red-green double	3 T
M 92		Hercules	between π Herculis & β Draconis	6.1	globular cluster	1 B; 2 T
ε Lyrae	the double-double	Lyra	left of Vega	4.6–4.9	pair of tw° double stars	4 B; 1 T
M 57	Ring neb.	Lyra	between β & γ Lyrae	9.3	planetary nebula	3 T
M 27	Dumb-bell	Vulpecula	between β Cygni and β Delphini	7.6	planetary nebula	2 B; 4 T
M 8	Lagoon	Sagittarius	5° west of λ Sagittarii	5.9	diffuse nebula	4 B; 4 T
M 17	Omega	Sagittarius	5° north east of μ Sagittarii	7.7	diffuse nebula	3 T
M 22		Sagittarius	3° west of Aguila	6.1	globular cluster	3 T
M 20	Triffid	Sagittarius	2° southwest of μ Sagittari	7.5	diffuse nebula	2 B; 3 T
M 23		Sagittarius	5° northwest of μ Sagittarii	6.9	open cluster	2 B; 3 T
M 6		Scorpius	near γ Sagittarii	5.3	open cluster	3 B; 1 T
M 7		Scorpius	5° northeast of λ Scorpii	3.2	open cluster	3 B; 1 T
M 5		Scorpius	Forms triangle with ε and μ Serpens	6.2	globular cluster	3 T

The numerical scale from 1 to 5 = very difficult and low interest to very easy and interesting to observe. Note that galaxies require exceptionally clear and transparent skies to be visible from the city.

B = Binoculars

T = Telescope

The autumn sky

*A*s fall approaches and the days shorten, astronomers look forward to more observing opportunities. The rain and more wind also help clear the air of urban pollution and thereby improve the quality of seeing and transparency.

A "w" shaped key to the fall sky

The longer nights and generally pleasant fall weather make this the best time of year to observe for most amateur astronomers. Although autumn skies do not offer as many bright stars as other seasons, they still hold a surprise or two in store for us.

Although still visible for several more weeks, the Great Summer Triangle is slowly setting in the west. The great "W" of Cassiopeia reins supreme at this time almost overhead and so serves as a handy celestial reference point. The central part of the W points in the direction of Polaris and the Big Dipper, now lost in the haze of the northern horizon.

The lower portion of the W directs our view toward a group of stars looking rather like a huge Big Dipper. This is the constellation Andromeda which is directly attached to the giant square of Pegasus, the mythical winged horse. Together these two constellations dominate the fall sky with their unique configuration.

The most distant object you can see

Here is an opportunity to make a little money with astronomy. Make a bet with a friend that you can see the most distant object visible to the human eye right from the city (with a little help from your binoculars, to be sure). Is it the tallest skyscraper or church? No. How about the Moon? No, again. It is M 31, the great galaxy in Andromeda at 2.2 million light

M 33 is very difficult to see from city locations. It is best left for a weekend trip under darker skies in the countryside.

years away. This galaxy, the closest major one in our galactic neighborhood, contains about 100 billion stars. It is easily located in the alignment from the stars ν and β Andromedae, and is just next to the latter. Look for it on a clear, moonless night, as an elongated nebulosity, and far more spectacular in binoculars than through a telescope.

It is sobering to realize that the light hitting your eyes at that instant, left this great galaxy at the dawn of humanity in the plains of Africa! M 31 is accompanied by M 32, a smaller, satellite galaxy that is very difficult to make out in a small telescope. Just south of β Andromedae, in a direction equal to but opposite of M 31, you will find Triangulum, a small, rather faint constellation. Try to locate M 33 just to the right of α Trianguli. This is

ABOUT FILTERS

Using (special interference) filters like "Deep Sky," UHC, Oxygen III, described on page 45, requires considerable familiarity with the sky. They are also quite expensive. A regular yellow filter can be as effective at much less cost in enhancing the relative contrast of many star clusters and nebulae. Although such a filter will not block out the yellow–orange light from high-pressure sodium lamps, it can enhance the overall contrast of several types of deep-sky objects.

Looking like diamonds against black velvet, the Double Cluster in Perseus is among the 25 top jewels in the sky.

another galaxy seen face on, but much fainter than M 31 and usually very hard to make out from the city.

A celestial jewel

Returning to Cassiopeia, look at the left "V" of the great W, pointing directly toward the constellation of Perseus. About half way between δ Cassiopeiae and α Persei (Mirfak) you will find the double cluster in Perseus, NGC 869 and 884. It is located about 7000 light years from us. Although its rather dull name does not indicate it, this cluster is one

of the most beautiful objects in the sky and extremely easy to find with binoculars. Calling it two sparkling diamonds on black velvet is no exaggeration.

The constellations of Cassiopeia, Perseus, and Cepheus, are all located in one of the outer arms of the Milky Way, opposite the center of our galaxy. Moving from Cepheus along this branch of the Milky Way with binoculars, we come to Auriga (a constellation better studied in the winter). This external region of our galaxy is remarkably beautiful.

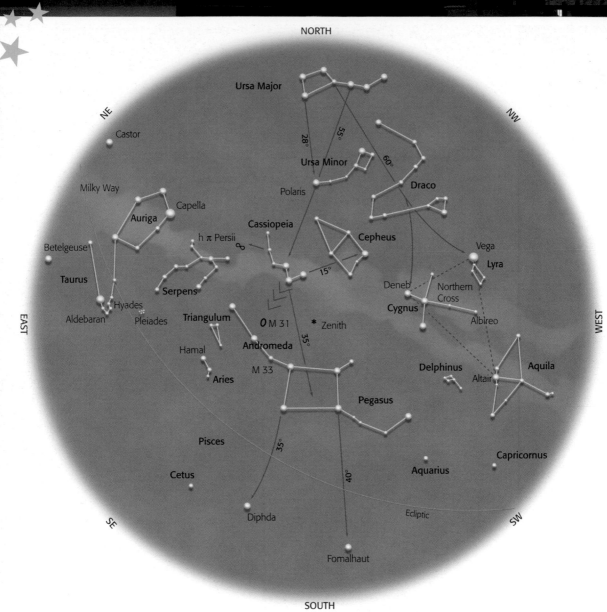

NORTH

Ursa Major

NE

Castor

Milky Way

Capella

Auriga

Cassiopeia

h π Persii

Betelgeuse

Taurus

Serpens

Hyades

Aldebaran   Pleiades

Triangulum

Hamal

Aries

Ursa Minor

Polaris

28°   55°

60°

Draco

Cepheus

15°

Vega

Lyra

Deneb   Northern
Cross

Cygnus

Albireo

Andromeda

O M 31   *Zenith*

M 33   35°

NW

Delphinus

Altair

Aquila

Pegasus

Pisces

Cetus

35°   40°

Aquarius

Capricornus

EAST   WEST

Diphda

Ecliptic

SE   SW

Fomalhaut

SOUTH

This fall sky chart can be used at the dates and
times indicated below. Note: all times given are in
universal time (UT) and you must compensate for
differences due to standard time
These charts apply on the following dates and
times:

- September 15 at 23h
- October 1 at 22h
- October 15 at 21h
- October 30 at 20h
- November 15 at 19h
- November 30 at 18h

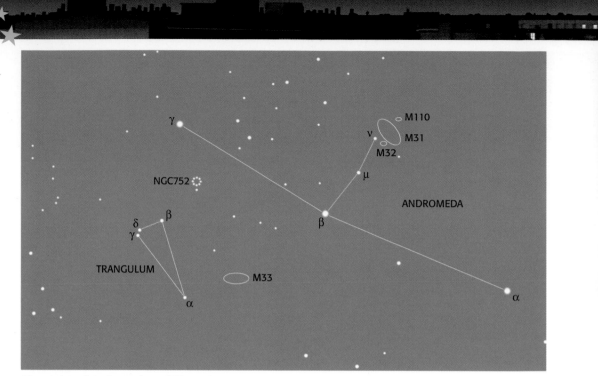

The region of Andromeda and Triangulum.

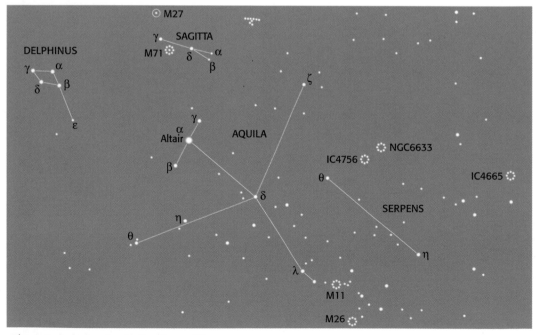

The area including Aquila, Sagitta and Serpens.

Interesting objects in the fall sky

Object	Name	Constellat.	Location	Mag.	Type of Object	Interest/ Difficulty
M 31	Andromeda Galaxy	Andromeda	along μ & ν Andromedae	3.8	galaxy	5 B; 5 T
γ Andromeda	Almach	Andromeda	along α & β Andromedae	2.3–5.1	yellow/green double star	4 T
M 32		Andromeda	next to M31	8.7	galaxy	1 T
γ Aries		Aries	1° south of β Aries	3.9	double star	2 T
NCG 457		Cassiopeia	2° south of δ Cassiopeiae	8.6	open cluster	3 B; 1 T
NGC 663		Cassiopeia	1° east of δ Cassiopeiae	8.4	open cluster	3 B; 3 T
M 52		Cassiopeia	between δ Cephei and χ Cassiopeiae	6.9	open cluster	1 B; 2 T
M 103		Cassiopeia	2° east of δ Cassiopeiae	7.4	open cluster	1 B; 1 T
M 15		Pegasus	near ε Pegasi	6.3	globular cluster	3 T
NCG 869 NGC 884	Double Cluster	Perseus	midway between γ Persei & δ Cassiopeiae	4.4–4.7	double open cluster	5 B; 5 T
M 34		Perseus	next to Algol (β Persei)	5.5	open cluster	1 B; 2 T
M 33		Triangulum	near α Trianguli	5.7	galaxy	1 B; 1 T
M 2		Aquarius	5° north of β Aquarii	6.3	globular cluster	2 T

The numerical scale from 1 to 5 = very difficult and low interest to very easy and interesting to observe. Note that galaxies require exceptionally clear and transparent skies to be visible from the city.

B = Binoculars

T = Telescope

The winter sky

*F*rosty nights are upon us and we are tempted to put away our telescopes until warmer weather returns. That would be a real pity, however, since the winter offers some of the finest sights in the sky. Often, too, those glacial winter nights have some spectacularly transparent skies and the heavens sparkle . . . so, give yourself that little extra push!

The richest part of the sky

When using the star charts on the following pages, start by identifying and linking the eight brightest stars you can see, Capella in Auriga, Castor and Pollux in Gemini, Procyon in Canis Minor, Sirius in Canis Major, Aldebaran, the eye of Taurus, and Betelgeuse and Rigel in Orion. At this time of year they form a giant "G" right in the middle of the sky.

The center of this G is dominated by the constellation of Orion and is the key to the rest of the winter sky. This great hunter of mythology forms a quadrilateral in the middle of the sky, with Canis Major at his feet and in pursuit of Lepus, the hare, all under the watchful red eye of Taurus, the bull. This grouping is overseen by Auriga, the charioteer, located just above our heads.

In the center of Orion, you can see an alignment of three remarkable stars, the three kings: Alnitak (ζ), Alnilam (ε) and Mintaka (δ), which together represent the hunter's belt. Just beneath it hangs Orion's sword, home to one of the most beautiful objects in the heavens, M 42, the Orion Nebula. It is visible to the naked eye even in the middle of the city. Study it carefully with binoculars or under low power with a telescope. With a 200-mm (8-inch) telescope, it takes on the shape of a greenish bird wing. The heart of the nebula is home to the Trapezium, a quartet of stars known as θ 1 and θ 2. This region of M 42 is an active stellar nursery that contains many very young stars, no more than a few

M 42, one of the 25 showcase objects in the sky. This immense cloud of hydrogen (and dust) is a veritable stellar hatchery.

million years old. The trapezium is easily resolved in 10×50 binoculars or in a small telescope.

Orion is also one of the few constellations that contain two prominent super giants, Betelgeuse and Rigel. The latter is also an orange and blue double star. The spectacular Horse Head nebula IC 434 is located just beneath Altinak (ζ). This dark nebula is an immense cloud of dust that obscures much of the large emission nebula behind it. Unfortunately, the Horse Head is strictly a photographic object and can only be glimpsed (under very dark skies) in large telescopes.

The brightest star in the sky

Extending the line of the three stars in Orion's belt to the left we come across Sirius, which at magnitude -1.4 is the brightest star in the sky. Sirius was one of the navigational reference points used by the Apollo moon missions, and of course by sailors and mariners well before that (certainly well before the invention of GPS!).

To the right from Orion's belt, we come across Aldebaran and the Hyades, a prominent group of stars at the center of Taurus and an open cluster just visible to the eye under good conditions. Moving fifteen degrees further west in the same direction, we come across another celestial showpiece, the Pleiades, also

known as the seven sisters. Most people can discern the seven main stars of this open cluster by eye and also notice that they form a tiny dipper-like asterism. While a treat for the eye, this magnificent cluster of young stars is far better viewed in binoculars than in a telescope. About 12° further east we find ζ Taurii, very close to the supernova remnant M 1. Also known as the Crab Nebula, this object is unfortunately very difficult to see from city skies.

Let us move further north to the center of the great constellation Auriga. This is a region rich in open clusters, including NGC 1746, M 37, M 36, and M 38, all nicely visible in binoculars and small telescopes. Just west of Capella (α Aurigae), we see ε Aurigae, one of the most luminous and largest stars in the galaxy, so large in fact that our entire solar system could fit within its confines!

As we near the end of this little book, let us quote the well-known, 19th-century, French writer and amateur astronomer, Camille Flammarion: "There is little doubt that astronomy engages the highest levels of the human intellect and spirit." That is probably truer today than it was at the time of Flammarion, who could surely have had no notion of humans walking on the Moon or instruments landing on Mars. His words still resonate today, however, especially as we look at the stars through out telescopes on a crisp winter night.

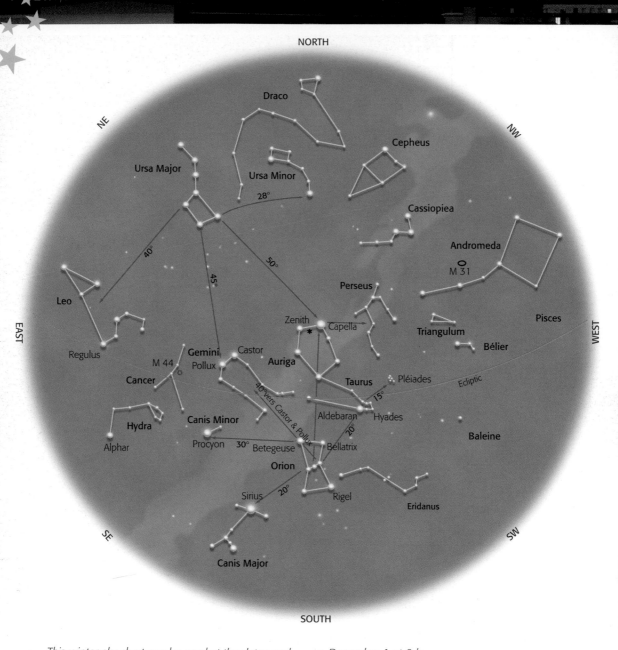

NORTH

Draco

Cepheus

NE

Ursa Major

Ursa Minor

28°

Cassiopiea

NW

Andromeda

M 31

40°

45°

50°

Perseus

Leo

Zenith

Capella

Triangulum

Pisces

Regulus

EAST

Gemini

Castor

Pollux

Auriga

Taurus

Pléiades

Bélier

Ecliptic

WEST

M 44

Cancer

15°

Hydra

Canis Minor

Aldebaran

Hyades

Baleine

Alphar

Procyon

30°

Betegeuse

Bellatrix

20°

Orion

Sirius

20°

Rigel

Eridanus

SE

Canis Major

SW

SOUTH

40°vers Castor & Pollux

This winter sky chart can be used at the dates and times indicated below. Note: all times given are in universal time (UT) and you must compensate for differences due to standard time.

These charts apply on the following dates and times:

- *December 1 at 0 h*
- *December 15 at 23 h*
- *January 1 at 22 h*
- *January 15 at 21 h*
- *January 31 at 20 h*
- *February 15 at 19 h*

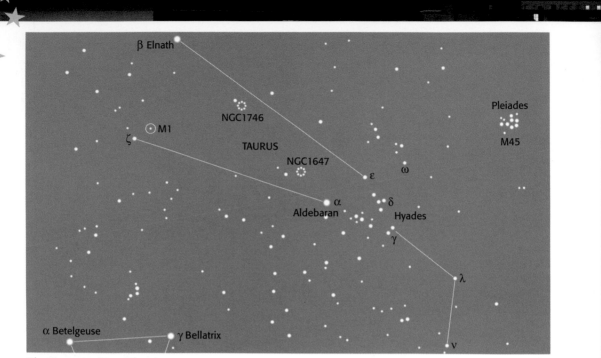

The Taurus region of the sky.

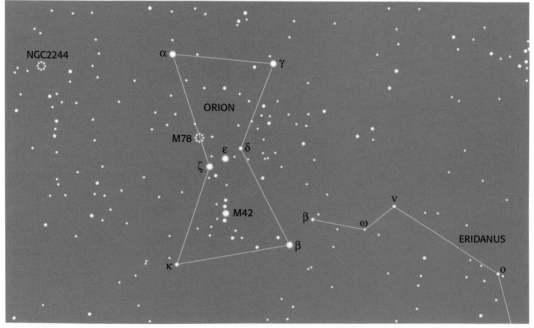

The region around Orion.

Interesting objects in the winter sky

Object	Name	Constella.	Location	Mag.	Type of object	Interest/ Difficulty
M 36		Auriga	between M 37 & 38	5.3	open cluster	1 B; 2 T
M 37		Auriga	between β & θ Aurigae	6.2	open cluster	1 B; 2 T
M 38		Auriga	between θ Geminorum & ι Aurigae	6	open cluster	1 B; 2 T
M 35		Gemini	between η Geminorum θ Aurigae	6.2	open cluster	3 B; 3 T
α Gemin.	Castor	Gemini	below Pollux	2	white/orange triple star	2 T
M 41		Canis Major	4° south of Sirius	4.6	open cluster	3 B; 3 T
M 42	Orion Nebula	Orion	under the belt	5	diffuse nebula	5 B; 5 T
σ Orionis		Orion	under Alnitak (ζ)	3.8	multiple stars	3 T
β Orionis	Rigel	Orion	southwest under the belt	0.2	blue/white doublet	3 T
δ Orionis	Mintaka	Orion	3rd belt star	2.3	double star	3 B; 3 T
M 78		Orion	under ζ Orionis near Betelgeuse	8	diffuse nebula	2 T
θ1 Orionis	Trapezium	Orion	center of M42	5.8 to 8	multiple stars	5 B; 5 T
M 45	Pleiades	Taurus	northwest of Aldebaran	1.4	open cluster	5 B; 5 T
M 1	Crab Neb.	Taurus	1° northwest of ζ Tauri	8.4	supernova remnant	1 T
	Hyades	Taurus	near Aldebaran		very open cluster	5 B; 2 T

The numerical scale from 1 to 5 = very difficult and low interest to very easy and interesting to observe. Note that galaxies require exceptionally clear and transparent skies to be visible from the city.

B = Binoculars

T = Telescope

Helpful information for the urban astronomer

The following pages contain an assortment of tables, references, hints, a bibliography and useful addresses and websites, most amateurs will find helpful as they progress in their hobby.

The Greek alphabet is part of the language of astronomy

In more ancient times, the stars were named in accordance with their location and position in the various constellations, the eye of the Bull, the left foot of Orion, the spear of Hercules, and so forth. Hardly very practical. Finally in 1603 the German astronomer Bayer introduced a much simpler way to name all stars using the Greek alphabet. This provided a universally accepted designation for stars while still retaining the 9th century Arabic names assigned to the most prominent stars.

THE GREEK ALPHABET	
α	alpha
β	beta
γ	gamma
δ	delta
ε	epsilon
ζ	dzeta
η	eta
θ	theta
ι	iota
κ	kappa
λ	lambda
μ	mu
ν	nu
ξ	ksi
ο	omicron
π	pi
ρ	rho
σ	sigma
τ	tau
υ	upsilon
φ	phi
χ	khi
ψ	psi
ω	omega

Principal meteor showers

Time of year	Shower name	Radiant constellation
Jan. 1–6 (peak on the 4th)	Quadrantids	Bootes
April 19–24 (peak on the 22nd)	Lyrids	Hercules/Lyra
May 1–13 (peak on the 5th)	Aquarids	Aquarius
July 25–30 (peak on the 28th)	Aquarids	Aquarius
Aug. 9–14 (peak on the 12th)	Perseids	Perseus
Oct. 9	Draconids	Draco
Oct. 16–22 (peak on the 19th)	Orionids	Orion
Nov. 17–23	Andromedides	Leo
Dec. 9–13 (peak on the 12)	Geminids	Gemini

Note that nothing in particular happens at the radiant, it is simply the apparent point of origin of the meteor shower. Meteors appear most frequently near and around this point in the sky. These showers recur on the same dates each year.

The 25 brightest stars

Number	Name of star	Constellation	Magnitude
1	Sirius	Canis Major	−1.4
2	Canopus	Carina	−0.7
3	Rigil Kentarus	Centaurus	−0.3
4	Arcturus	Bootes	−0.06
5	Vega	Lyra	0
6	Capella	Auriga	0.1
7	Rigel	Orion	0.1
8	Procyon	Canis Minor	0.4
9	Achemar	Eridanus	0.5
10	Agena	Centaurus	0.6
11	Altair	Aquila	0.7
12	Betelgeuse	Orion	0.8
13	Aldebaran	Taurus	0.8
14	Antares	Scorpius	0.9
15	Acrux	Southern Cross	0.9
16	Spica	Virgo	1.0
17	Pollux	Gemini	1.1
18	Formalhaut	Pisces Australis	1.1
19	Regulus	Leo	1.3
20	Deneb	Cygnus	1.3
21	Mimosa	Southern Cross	1.3
22	Adhara	Canis Major	1.5
23	Castor	Gemini	1.6
24	Schaula	Scorpius	1.6
25	Bellatrix	Orion	1.6

Lunar eclipses to 2010

Date	Type	Duration (hours)	Maximum (UT)	Visibility (Europe)	Visiblity (N. America)
May 16, 2003	Partial	0.49	3 h 43	Beginning visible	Entire continent
Nov. 9, 2003	Total	1.13	1 h 20	Visible	Entire except west coast
May 4, 2004	Total	1.30	20 h 30	End visible	Invisible
Oct. 28, 2004	Total	1.31	3 h 04	Visible	Entire continent
Oct. 17, 2005	Partial	0.06	12 h 02	Invisible	Entire continent
Sept. 7, 2006	Partial	0.18	18 h 54	End visible	Invisible
March 3, 2007	Total	1.23	23 h 23	Visible	Total for Great Lakes & eastward. Partial for rest of continent
Aug. 28, 2007	Total	1.47	10 h 36	Invisible	Total for Rocky Mountain region and westward
Feb. 21, 2008	Total	1.11	3 h 28	Visible	Entire continent
Aug. 16, 2008	Partial	0.81	21 h 09	End visible	Beginning visible
Dec. 31, 2009	Partial	0.07	19 h 25	Visible	Invisible
June 26, 2010	Partial	0.53	11 h 39	Invisible	Invisible
Dec. 21, 2010	Total	1.25	8 h 17	Beginning visible	Entire continent

Basic data on the planets

	Diameter (Km)	Mass (Earth=1)	Distance from Sun (millions of Km)	Period of revolution (in days)	Rot'n. Period (days)	Average Temp	# of satellites	Orbital Inclination to ecliptic
Mercury	4 878	0.055	57.6	87.96	59	400° C	0	7° C
Venus	12 104	0.82	108.2	224.7	243	450° C	0	3°23'24"
Earth	12 756	1.0	149.6	365.25	23h56'18"	19° C	1	0°0'
Mars	6 785	0.107	227.9	686.98	24h 37'	0° C	2	1°51'
Jupiter	142 980	317.89	778.3	11.9 yrs*	9h 50'	−100° C	16 (+?)	1°18'36"
Saturn	120 540	95.18	1 429.4	29.5 yrs*	10h 39'	−180° C	18 (+?)	2°30'
Uranus	51 120	14.53	2 875	83.75yrs*	17h 20'	−220° C	18	0°46'12"
Neptune	49 530	17.15	4 504.4	163.7 yrs*	15h 50'	−220° C	8	1°46'12"
Pluto	2 300	0.002	5 915.8	248 yrs*	6d 9h 18'	−240° C	1	17°10'12"

* Terrestrial days

Note that an Earth day is not 24 h long, but 23h 56' 18", which corresponds to one sidereal rotation. However, it takes 24 h on average for a given point on Earth to come around fully facing the Sun (midday). This 3 min 42 s difference is due to the Earth's orbital motion.

The 25 most beautiful objects in the sky

This list of 25 carefully selected deep-sky objects includes some of the best-known and spectacular astronomical sights that everyone should become familiar with. Some of them, particularly the galaxies, can really be fully appreciated only under dark country skies away from urban light pollution. Save those for your next vacation.

Designation	Description	Constellation	Binoculars or Telescope?
M 31/ M 32	The brightest galaxy, naked eye object	Andromeda	Both
M 44	The Praesepe (occasionally naked eye)	Cancer	Both + NE
M 51	Whirlpool galaxy, one of the brightest	Canis Venatici	Both
Alberio	One of the most colorful double stars	Cygnus	Telescope
M 11	Splendid open cluster	Scutum	Both
M 41	Grand open cluster near Sirius	Canis Major	Both
M 81 & M 82	A pair of bright galaxies	Ursa Major	Both
Mizar	Well-know double near Alcor	Ursa Major	Telescope
M 13	The best-known globular cluster	Hercules	Both
M 46 & M 47	A double open cluster	Puppis	Both
Algieba	Beautiful colored double star	Leo	Telescope
M 57	The Ring Nebula	Lyra	Telescope
ε1 & ε2 Lyrae	The famous double-double	Lyra	Telescope
M 42	The Orion Nebula, the brightest nebula and visible to the naked eye	Orion	Both + NE*
Theta 1 (θ) Orionis	The Trapezium, quadruple star at the center of the Orion Nebula	Orion	Telescope
h & χ Persei	The famous Double Cluster	Perseus	Both + NE*
M 27	Dumb-Bell Nebula, bright planetary	Vulpecula	Both
M 22	Bright globular cluster	Sagittarius	Both
M 8	Lagoon Nebula, brightest after M 42	Sagittarius	Both
M 20	Triffid Nebula	Sagittarius	Both
M 4	Globular cluster near Antares	Scorpius	Both
M 7	Beautiful open cluster	Scorpius	Binoculars + NE*
M 45	The Pleiades, splendid open cluster	Taurus	Both + NE*

NE = Naked eye object

What you can see with binoculars

	Binoculars 7×50–10×50	Small Refractor 50–60 mm at 20× to 50×
Sun (fully filtered)	Not really suitable for solar viewing	Sunspots and their daily motion across solar disk
Moon	Major basins, crater rays, Earthshine	Mountain shadows, major craters, basin detail
Jupiter	Disk of planet and 4 Galilean moons	Flattened disk, movement of moons
Saturn	Looks like a "fat star"	Rings clearly visible when wide open
Venus	Very occasionally the crescent at 10×	Crescent during inf. conjunction
Mars	Looks like an orange star	Small orange disk
Other Solar system objects	Finding comets and their general aspect, finding Mercury	Finding comets and study in detail
Star and deep sky objects	Milky Way stars, Pleiades, Hyades, Praesepe, Orion Nebula, M 31 & bright diffuse objects like M 16	Open and globular clusters, double stars with at least 10" separation

Monthly planetary positions can be found in ephemeredes and many astronomy magazines and are usually shown on charts indicating the constellations along the ecliptic in which the planets are located at any given time. For details and positions of star clusters, nebulae, and galaxies, refer to pages 72 to 93 in this book.

and various size telescopes

70–80 mm Refractor at 100×–120×	100–120 mm telescope (4–5") at 150×–200×
Sunspot details: umbra, penumbra, faculae	Daily changes in sunspot detail and filaments
Major clefts, crater detail, Straight Wall, basin 'waves'	Very fine detail in craters like Plato, soft coloration
Parallel cloud bands, transits of moons, color	Fine detail, rotation timings (difficult), Red Spot
Main moon Titan, Cassini's division in rings, flattened shape of disk	3–4 moons (Titan, Iapetus ,Tethys, Rhea). Shadow of rings on planet and occasional marking
Follow all phases, but no surface detail	Occasional dusky cloud markings and phase irregularities
White polar caps, some dark marking when at opposition	Major dark features (Syrtis Major, Sinus Meridiani, Mare Acidalium), planet's rotation
Phases of Mercury, disk of Uranus, low power views of comets	Very rare detail on Mercury, small blue –green disk of Neptune (Pluto is not visible ever)
Omega Nebula, Triffid, Lagoon, M 78 and M 27. Double stars with at least 5" separation	Galaxies M 33, M 51 (in Vulpecula) and M81. Planetary nebulae: Dumb-Bell (M 27), Ring (M 57), Double stars to 2" separation

Be sure to look for these dim objects on dark nights with good transparency and no Moon. Be sure as well to wait 15 minutes or so to become fully dark-adapted. Deep-sky objects can be disappointing in small instruments, so use as low power as possible to observe them. This will give you the widest field of view and the brightest images possible.

Cleaning and maintenance

*M*ost beginners are eager to use their new equipment as soon as possible and often hesitate to make those all-important "fine tuning" adjustments. That is a mistake. Even though what you see through a new telescope seems entirely acceptable at first glance, there is often room for improvement. Whether you are using binoculars or working with a telescope, proper maintenance and adjustment is essential to get the most out of the equipment.

Adjusting binoculars

Proper adjustment of binoculars is done in two stages. First, you must adjust the angle and inter-pupillary distance of the two eyepieces to match those of your own eyes. You can do this easily by looking at the sky or ceiling in daylight and moving the two hinged binocular stems right or left until you see but a single image. Most people have an inter-pupillary distance of about 65 mm (although there is considerable individual variation in this). If this distance is not properly adjusted you risk developing a headache.

A second important adjustment is correction for diopter differences between your own two eyes. To do this:

Since our individual two eyes are rarely identical, binoculars are equipped with separate diopter adjustment knobs to compensate for this difference.

- Cover the right objective with its dust cap, and focus with your left eye on a contrasting nearby object like a wall or some foliage.
- Next uncover the right objective and place the dust cap on the left. Now focus again on the same object by lightly turning the diopter ring on the right side eyepiece (which is normally preset at 0).
- Remove the objective dust cap and look through the binoculars with both eyes to ensure the image is sharp and your view is comfortable.
- Diopter adjustment of plus or minus 4 are possible, depending on how far or nearsighted you may be.

- Remember your personal diopter value so that you can pre-adjust your binoculars when you use them again.

Note that if you have a pair of roof prism binoculars, the diopter adjustment knob is usually located in the central stem close to the main focusing knob. The operating principle is the same however.

Adjusting a telescope finder

After centering your target in the main telescope, adjust the finder by simultaneously turning two setscrews until the cross hairs are also centered on the object.

Although many beginners are initially reluctant to tamper with their telescope finders, adjusting them is actually very simple and really essential for effective use of your instrument. Finders tend to get knocked out of alignment quite often and you will have to re-center them on a regular basis. If your finder is mounted by two rings with three set screws each (3 in front and 3 in back), all you have to do is adjust either the front or back set and leave the other locked in place. Start by adjusting the set screws in one ring so that the finder is roughly centered in it, and use the other three screws to point your finder as needed.

- Be sure to partially tighten the lock knobs on the telescope mount first to keep it from moving around as you adjust the finder.
- Place a low power eyepiece in the telescope and aim it like a rifle at a distant a TV antenna or a lamppost, and center it in the field of view. The "T" shape of an antenna is handy for this.
- Next, look through the finder and center its cross hairs on the same object. You will need both hands for this. Gently turn one of the finder set screws in one direction and another one in the opposite direction, making sure there is no slop in this motion. As you continue this be sure to always turn two set screws at the same time. Note the direction of image shift and try to keep it as close as possible to the finder cross hairs. Keep doing this until the object is as closely centered in the finder as possible.
- Once you are satisfied with that, use a medium power eyepiece in the telescope and repeat as above. Continue this until the finder and telescope images are perfectly centered at the highest magnification you intend using. Once the finder is fully centered, gently lock all sets screws in place.

Maintaining your equipment

Binoculars and telescopes usually require little in the way of maintenance, but do need some basic care. Above all, avoid dust as much as possible. Cap or package your optics after each use and keep everything well covered. If you

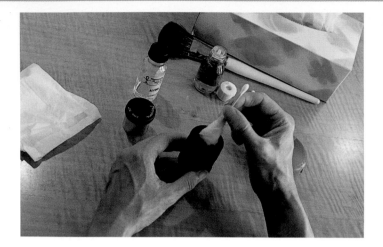

An assortment of tools to keep your eyepieces clean. While cleaning them, be sure to never rub the surfaces hard or with much pressure. The eye lenses of your oculars are usually the ones that get soiled by the observer's eyelashes.

are planning an extended break during an observing session, cover or cap the telescope tube temporarily.

The best way to keep your optics clean is to not get them dirty in the first place! Avoid getting fingerprints on objectives lenses and eyepieces, especially those of curious little hands eager to explore.

Keep in mind that some dust on the front lens or corrector plate of your telescope will not appreciably affect image quality. You should only have to clean them occasionally, once or twice a year at most. If you own a Newtonian (or Dobsonian), we suggest you have the mirror cleaned by experts only and do not try to dismantle it yourself.

The best way to clean an objective lens or corrector plate is to first blow away as much dust and grit as possible with a photographer's rubber bulb. Never use compressed air in a can. To remove really tenacious dust and dirt use a camel hairbrush or soft make-up brush. Once you are sure all grit has been removed,

gently wipe (without RUBBING) the optical surface with a large, makeup style cotton ball wrapped in (unscented and lotion free) Kleenex or similar white tissue paper. While doing this, keep the optical surface slightly damp with your breath and do not apply any pressure while wiping. If there is any residue left, get fresh tissue paper and continue wiping. Any persistent finger prints can be removed with tissue or cotton dampened with optical cleaning fluid.

Cleaning eyepieces

The observer's moving eyelashes inevitably smear ocular eye lenses. First blow off any dust with a rubber bulb and then use cotton tipped swabs dampened with lens cleaning fluid. Wipe the lens surface gently and the dry it with a clean cotton swab. Always store your eyepieces in their original boxes and NEVER take them apart to clean the inside lenses.

Bibliographic and reference material suggestions

Astronomy magazines and associated websites

Astronomy (www.astronomy.com)

Sky & Telescope (skyandtelescope.com)

Astronomy Now: this is Britain's most popular astronomy magazine (www.astronomynow.com)

Sky News: top Canadian publication with much US readership as well (www.skynewsmagazine.com)

Sky & Space: top Australian/New Zealand publication (PO Box 1233, 80 Elby Street, Bondi Junction, NSW 1355, Australia)

Journal of the Association of Lunar and Planetary Observers: published by the association of the same name. Particularly useful for urban observers specializing in solar system observing. (www.lpl.arizona.edu/alpo)

Introductory books

In addition to a huge selection from Cambridge University Press and Sky Publishing Corp., the following have been extremely popular with US and Canadian amateurs:

NightWatch : by Terence Dickinson, 3rd edition, Firefly Books, 1998

The Backyard Astronomer's Guide: by Terence Dickinson and Alan Dyer, Revised edition, 2002, Firefly Books.

General reference books

Observer's Handbook: Published annually by the Royal Astronomical Society of Canada

Handbook of the British Astronomical Association: Published annually by the BAA

Astronomical equipment and suppliers

In addition to **Meade** and **Celestron** in the USA, some of the most popular and long-established suppliers of affordable, quality equipment include:

ORION Telescopes and Binoculars: PO Box 1815, Santa Cruz, CA 95061 (www.telescope.com)

Astronomics: 680 SW 24th Ave., Norman, OK 73069, USA (www.astronomics.com)

Index

Photographic credits

Toutes les photographies de l'ouvrage sont de Denis Berthier, sauf :

p. 7 © Alain Cirou/Ciel&Espace –
p. 8-9 © E.Graëff/Ciel&Espace – p. 18 © NASA –
p. 19 © Christian Arsidi – p. 17 © Televue –
p. 20-21 © Celestron – p. 28-29 © E. Graëff/
Ciel&Espace – p. 32 © Unterlinden –
p. 36-37 © SPJP – p. 40 © SPJP – p. 41 bas © SPJP –
p. 42 gauche © Unterlinden, droite © Médas, App'ar
studio Vichy – p. 43 haut gauche © SPJP, haut droite

© Televue, bas droite © Unterlinden –
p. 50 © Gérard Thérin – p. 51 droite © Gérard
Thérin – p. 52-53 DR – p. 55 © Marc Larguier –
p. 56 bas © F. Espenak/Ciel&Espace –
p. 58 © G. et Y. Delaye – p. 63 © Serge Brunier –
p. 64 © Serge Brunier – p. 65 © NASA –
p. 70 © Viladrich – p. 76 © Celestron –
p. 83 © Gérard Thérin